DEMONIC
MALES

EDITED BY RICHARD WRANGHAM

Chimpanzee Cultures
with William C. McGrew, Frans B. M. de Waal,
and Paul G. Heltne

Primate Societies
with Barbara B. Smuts, Dorothy L. Cheney,
Robert M. Seyfarth, and Thomas T. Struhsaker

Ecological Aspects of Social Evolution:
Birds and Mammals
with Daniel I. Rubenstein

Current Problems in Sociobiology
with Brian C. R. Bertram, Timothy H. Clutton-Brock,
Robin I. M. Dunbar, and Daniel I. Rubenstein

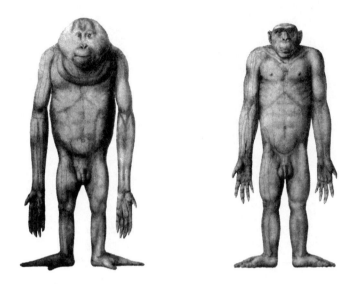

RICHARD WRANGHAM
AND DALE PETERSON

HOUGHTON MIFFLIN COMPANY
BOSTON   NEW YORK

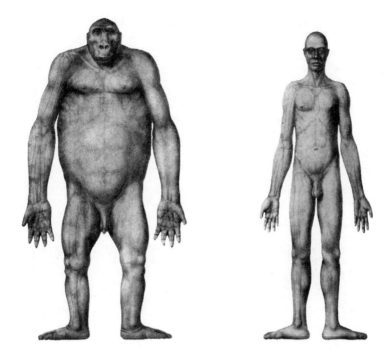

DEMONIC MALES

APES AND THE ORIGINS
OF HUMAN VIOLENCE

For information about permission to reproduce selections from
this book, write to Permissions, Houghton Mifflin Company,
215 Park Avenue South, New York, New York 10003.

For information about this and other Houghton Mifflin
trade and reference books and multimedia products, visit
The Bookstore at Houghton Mifflin on the World Wide Web
at http://www.hmco.com/trade/.

Library of Congress Cataloging-in-Publication Data
Wrangham, Richard W.
  Demonic males : apes and the origins of human violence /
Richard Wrangham and Dale Peterson.
    p. cm.
  Includes bibliographical references and index.
  ISBN 0-395-69001-3
  1. Apes — Behavior. 2. Aggressive behavior in animals.
I. Peterson, Dale. II. Title.
QL737.P96W73   1996
156'.5 — dc20   96-20346   CIP

Printed in the United States of America

QUM 10 9 8 7 6 5 4 3 2

Book design by Melodie Wertelet

Maps by Mary Reilly

CREDITS

Title page illustration: "The body proportions of adult male orang-utan, chimpanzee, gorilla, and human, all reduced by the same scale and drawn . . . with the aid of accurate measurements of actual bodies. All hair has been omitted and the lower limbs are shown unnaturally straightened to facilitate comparisons." From *The Life of Primates* by Adolph H. Schultz. Weidenfeld and Nicolson, 1969. Reproduced by permission of the Orion Publishing Group Ltd.

The following drawings are by Mark Maglio and were reproduced by permission of the artist: page 9: Frodo, a Gombe chimpanzee; page 149: Male and female gorilla pair; page 207: Mother and offspring bonobos; page 229: A male bonobo.

The foot drawings in the Ape Evolutionary Trees on page 261 are based on images in *The Life of Primates* by Adolph H. Schultz (Weidenfeld and Nicolson, 1969); Karen Kilmer of the San Diego Zoo provided information about the bonobo foot. Reproduced by permission of the Orion Publishing Group Ltd.

The Ape Distribution Map on page 262 is based on maps featured in Jaclyn H. Wolfheim's *Primates of the World: Distribution, Abundance, and Conservation.* University of Washington Press, 1982. Reproduced by permission.

Dedicated to the memory of

two who found another way —

Geoffrey W. Wrangham

Paul G. Peterson

CONTENTS

1 / Paradise Lost   1

2 / Time Machine   28

3 / Roots   49

4 / Raiding   63

5 / Paradise Imagined   83

6 / A Question of Temperament   108

7 / Relationship Violence   127

8 / The Price of Freedom   153

9 / Legacies   173

10 / The Gentle Ape   200

11 / Message from the Southern Forests   220

12 / Taming the Demon   231

13 / Kakama's Doll   252

Family Trees   261

Maps   262

Notes   265

Bibliography   301

Acknowledgments   331

Index   333

DEMONIC
MALES

1

PARADISE LOST

"**Y**OU WILL BE KILLED!" the man at the Burundian embassy in Kampala said, in a bizarrely cheerful voice, as he stamped our visas.

But killing was the reason we were in Africa. Dale Peterson and I were exploring the deep origins of human violence, back to the time before our species diverged from rainforest apes, 5 to 6 million years ago. Not only ancestral to humans, those early rainforest apes were also part of a genetic line now represented by the four modern great ape species: orangutans, gorillas, chimpanzees, and bonobos. Both of us had already observed orangutans in Borneo and gorillas and chimpanzees in Africa, but neither of us had yet seen the fourth and rarest ape, the bonobo, in the wild.

To get to the bonobos, we first had to reach Bukavu, a town on the eastern side of Zaïre, just across the border from Rwanda. In Bukavu, we would pick up a single-engine plane and fly west for three hours across a sea of forest until, having passed more than halfway across the continent, we would find an airstrip and a little town isolated in that great green world. Near the town was the small pocket of rainforest where the bonobos lived.

To fly directly from Uganda to Zaïre was impossible because the shaky Zaïrean government, fighting for control of the country, had closed all international airports, and driving overland was not advised because of discouraging reports about bandits and guerrillas. And so we had decided to fly south from Uganda into Burundi, then drive a rented van through Burundi and Rwanda, and on into eastern Zaïre.

But Burundi was not at peace either. Half a dozen waves of ethnic slaughter had swept that small nation in recent years. In one month in 1972, Tutsis killed nearly every Hutu leader and any other Hutu who appeared literate. Tutsis, though they accounted for only about 15 percent of the populace, thus during the next two decades controlled the civil service, the military, and, within a one-party system, the upper reaches of government. Until 1993, when the country experimented with democratic, multiparty elections for the first time in history, every president had come from the minority Tutsis.

The June 1993 elections produced Burundi's first Hutu president, a political moderate, a believer in nonviolence and ethnic reconciliation, Melchior Ndadaye. But in the early morning of October 21, 1993, four months before we paid for our visas, an army tank rammed a hole in the white stucco wall of the presidential palace and radical Tutsi soldiers stabbed President Ndadaye to death with a knife. They also assassinated a half dozen other high officials from Ndadaye's government; the surviving ministers took refuge behind French troops at a hotel in Bujumbura, Burundi's capital city.

As the surviving ministers broadcast over Radio Rwanda appeals for the people to "rise up as one man and defend Burundi's democratic institutions,"[1] Hutus around the country took rough weapons, mostly machetes and spears, and slaughtered Tutsis over the next three months. In return, Tutsi soldiers and civilians massacred Hutus whenever possible.

The airport of Bujumbura was quiet, just about empty, and watched over by men with guns when we flew in from Kampala on February 12. Someone said (in French): "The road is good today. It wasn't yesterday. May not be tomorrow." So we jumped into a rented van and drove through the lowlands west toward Rwanda and Zaïre.

Burundi was green and cool and damp. We crossed a fertile land with rippled lakes of grass and corrugated fields growing maize and manioc. There were herds of long-horned cattle, rich smells, women carrying long bundles of twisted sticks on their heads, a woman wrapped in cloth standing in a fresh field and lifting and dropping a hoe. Unfriendly men in uniform, holding guns, stopped us at a roadblock, looked at our papers, and then allowed us to pass.

A few hours and three roadblocks later, we passed through immigration and customs into Rwanda. The road curved into the hills and then into the mountains and began disintegrating, yet still turning and winding ahead and up into a promising place where cloud and mountain casually intermingled. We stopped for a moment, looked over a sweep of floodplain and toothy mountains rising in the distance, listened to the roar and rush of a river below, and then, back in the van, continued moving through this elevated paradise of villages and small plantation plots edged by banana trees and bamboo fences.

The troubles hadn't come to Rwanda yet. They wouldn't come for another seven weeks, until April 6, when the Rwandan president and Burundi's interim president were assassinated. The presidents were returning together from a conference in Tanzania when their descending plane was shot down over the capital of Rwanda by unidentified men firing from the ground.

In a mirror image of the situation in Burundi, Rwandan Hutus controlled the army and the government while Tutsis were kept out. Within the first three days following the assassination, the Hutu army and militia began to carry out a well-organized campaign of genocide. The army executed all opposition leaders: sixty-eight Tutsis and moderate Hutus. Transitional Prime Minister Agathe Uwilingiyimana was murdered.

United Nations guards protecting her were tortured, sexually mutilated, and killed. The minister of labor was cut into three pieces and used as a roadblock.

Then the *real* killing began. Tutsi men, women, and children were massacred in the Red Cross refugee camps where they sought protection. Tutsi patients and staff were hacked to death in a hospital as foreign doctors watched. Tutsi families huddling for sanctuary inside a mission were blown up with hand grenades, then doused with gasoline and set on fire; the few survivors who tried to run away were cut down with machetes.[2] Estimates of the number dead ranged as high as half a million, whose blood and bodies literally flowed out of that small, beautiful country. Sweeping along the muddy Rusumo River into Tanzania, "piles of corpses bobbed like rag dolls," according to a reporter for *Newsweek*.[3] Authorities in Uganda estimated 10,000 bodies had washed down the Kagera River, out of Rwanda and into Lake Victoria, where they washed up against Ugandan shores.

Ngoga Murumba, a Ugandan farmer hired to haul the bodies out of the lake and dispose of them, described a numbing of horror, a blurring of memory. He had wrapped in plastic sheeting and stacked hundreds of bodies; only a single vision disturbed his mind. "One time I found a woman," he said. "She had five children tied to her. One on each arm. One on each leg. One on her back. She had no wounds. . . ."

At the end of Rwanda we descended, made a turn, and came to a body of water, a crowd of people, some cars, men with guns, a couple of barriers, and slid — after some minor harassment, negotiation, and small bribes solicited and rejected — into Zaïre.

The horror in Rwanda had not yet begun. That woman and her five children were still alive somewhere in the country. We proceeded across the border into Zaïre, then flew onward across hundreds of miles of rainforest into Equateur Province, to a landing strip in the village of Djolu. By following the lines of human ancestry back toward our common ancestry with the great apes, we were looking for patterns of behavior that would offer, so we believed, clues to a profound and disturbing mystery of the human species.

Some twenty years before our trip into central Zaïre, during the early afternoon of January 7, 1974, in Gombe National Park, Tanzania, a group of eight chimpanzees traveled purposefully southward, toward the border of their range. They were a good fighting party: seven males, six of them adults, one an adolescent. The alpha male, Figan, was there. So was his rival, Humphrey. The only female with them was Gigi. Childless and tough, she wouldn't slow them down.

As they walked, they heard beyond them calls from the neighboring community, but they didn't shout or scream in reply. Instead they maintained an unusual silence and quickened their pace. They reached their border zone, but they didn't stop. Soon they were beyond their normal range, moving very quietly into the neighbors' territory. Breathlessly keeping pace with them was Hillali Matama, the senior field assistant from Jane Goodall's research center in Gombe.

Just inside the neighboring territory, Godi ate peacefully, alone in a tree. Godi was an ordinary male: a young adult about twenty-one years old and a member of the Kahama chimpanzee community. There were six other males in Kahama, and those earlier calls had told him where some of his comrades were. Often they all traveled together. But today Godi had chosen to eat alone. A mistake.

By the time he saw the eight intruders they were already at his tree. He leapt and ran, but his pursuers raced after him, the front three side by side. Humphrey got to him first, grabbing a leg. Godi, unbalanced, toppled at once. Humphrey jumped on him. Leaning with his full weight of 50 kilograms, pinning his opponent like a wrestler, holding down two of his limbs, Humphrey immobilized him. Godi lay helpless, his face crushed into the dirt.

While Humphrey held, the other males attacked. They were

hugely excited, screaming and charging. Hugo, the eldest, bit Godi with teeth worn almost to the gums. The other adult males pummeled his shoulder blades and back. The adolescent watched from a distance. The female, Gigi, screamed and circled around the attack. (Imagine being battered by five heavyweight boxers and you have an idea of how Godi may have felt. Measured tests have demonstrated that even poorly conditioned captive chimpanzees are four to five times as strong as a human athlete in top condition.)[4]

After ten minutes Humphrey let go of Godi's legs. The others stopped hitting him. Godi lay face down in the mud while a great rock was hurled toward him. Then, still wild with excitement, the attackers hurried deeper into the Kahama territory, hooting and charging. After a few minutes they returned to the north and back across the border into their own range. And Godi, slowly raising himself, screaming with fear and anguish, watched his tormentors go. There were appalling wounds on his face, body, and limbs. He was heavily bruised. He bled from dozens of gashes, cuts, and punctures.

He was never seen again. He may have lived on for a few days, perhaps a week or two. But he surely died.

The attack on Godi was a first. Certainly not the first time that chimpanzees had made a raid into the neighboring range to attack an enemy — but the first time any human observer had watched them do it. It is the first recorded instance of lethal raiding among chimpanzees, and among chimpanzee observers and animal scientists in general, it struck a momentous chord.

This sort of thing wasn't supposed to happen among nonhumans. Until the attack on Godi, scientists treated the remarkable violence of humanity as something uniquely ours. To be sure, everyone knew that many animal species kill; but usually that killing is directed toward other species, toward prey. Individual animals — often males in sexual competition — fight with others of their own species; but that sort of contest typically ends the moment one competitor gives up. Scientists thought that only humans deliberately sought out and killed members of their own species. In our minds, we cloaked our own species' violence in culture and reason, two distinctly human attributes, and wondered what kind of original sin condemned us to this strange habit. And suddenly we found this event in the ape world. The attack on Godi suggested that chimpanzees might be a second species that killed its own kind deliberately. But how strange that the second species should be chimpanzees! After all, no species is more closely related to us than chimpanzees are.

What did it mean? Did Godi's pain point to a shared holdover from our evolutionary past? Did it imply that human killing is rooted in prehuman history? Or was Godi's death an aberration, a once-in-a-lifetime oddity, a meaningless expression of temporary ape insanity?

Time would tell.

Lethal raiding wasn't the only dark behavior to emerge during those early years of chimpanzee field research. Scientists also began to note instances of sexual violence.

When it comes to having sex, a female chimpanzee isn't normally very picky. She finds most males attractive, or at least tolerable. One kind of relationship, however, stops her in her tracks. She doesn't like to mate her maternal brothers. Even when those males court elaborately, with shaking branches and rude stares and proud postures, female chimpanzees refuse their brothers.

Normally, the female's reluctance to mate with her brother marks the end of it. But occasionally a brother can't stand being denied. She resists and avoids him. He becomes enraged. He chases and, using his greater size and strength, beats her. She screams and then rushes away and hides. He finds her and attacks again. He pounds and hits and holds her down, and there's nothing she can do. Out in the woods, there's a rape.

For many of us, our first pictures of chimpanzees in the wild were from Jane Goodall's gentle portraits. In the early 1960s, 5 million years after our species separated, Jane Goodall and

David Greybeard touched hands in mutual wonder, and a new kind of contact was established. Living in the wilds of Gombe, Goodall sketched for an entranced world the emotional lives of the apes she gave common names to: David Greybeard, Mike, Flo, Fifi, Gigi, and others. They had personalities we could understand: kind and gentle David, bold and daring Mike, lusty and savvy Flo. Their human gestures and vivid faces made them real as individuals. Their strong relationships made them seem familiar. We could relate to them and to what they do — extraordinary, humanlike things that were an eerie reminder of our shared ancestry.

Imagine that you're in chimpanzee country in West Africa, for example, walking through a warm, dark forest, and you hear a hammering sound. You follow it, thinking perhaps you're near an African village. Pressing through a thicket, you finally emerge into a relatively open area and see wild chimpanzees patiently working under a big nut tree, using stone hammers at their stone anvils, tap-tap-tapping on a hard nut until it breaks open. Sometimes the anvil is poorly balanced. The hammerer sees the problem, selects a smaller stone, and pushes it under the anvil as a stabilizing wedge. *Tap-tap-tap.* A youngster is trying, but she doesn't yet have the technique. The hammer doesn't seem to work. The mother takes it from her offspring, turns it upside down, and shows how it's done. A few minutes later, the daughter takes the stone back and tries her mother's way. *Tap-tap-tap.* Someone's smashed shell has a piece of nutmeat still tightly packed inside. She observes the problem, reaches for a twig, selects a spine, and uses it to pick out the last morsels.

Chimpanzee traditions ebb and flow, from community to community, across the continent of Africa. On any day of the year, somewhere chimpanzees are fishing for termites with stems gently wiggled into curling holes, or squeezing a wad of chewed leaves to get a quarter-cup of water from a narrow hole high up in a tree. Some will be gathering honey with a simple stick from a bee's nest, while others are collecting ants by luring them onto a peeled wand, then swiping them into their mouths.

There are chimpanzees in one place who protect themselves against thorny branches by sitting on leaf-cushions, and by using leafy sticks to act as sandals or gloves. Elsewhere are chimpanzees who traditionally drink by scooping water into a leaf-cup, and who use a leaf as a plate for food. There are chimpanzees using bone-picks to extract the last remnants of marrow from a monkey bone, others digging with stout sticks into mounds of ants or bees or termites, and still others using leaf-napkins to clean themselves or their babies. These are all local traditions, ways of solving problems that have somehow been learned, caught on, spread, and been passed across generations among the apes living in one community or a local group of communities but not beyond.[5]

When feeding is done, the apes relax. In a sunny glade in an otherwise deeply shaded forest, six chimpanzees sack out after the morning meal. The only mother lies on her side with her head on her curled arm. In front of her sits her grown-up son, handsome and erect, chin raised to allow his cousin and lifelong friend to clean his beard. But the mother's gaze is drawn across the glade to another male, still slim at eighteen years old, verging on adulthood. He lies with his eyes closed, dead to the world it would seem, except that with his right foot he is playing with her two-year-old daughter. Every time the daughter totters forward to grab his foot, it glides forward into her tummy and knocks her over. All is quiet, peaceful. The youngster's chuckly little laugh is the only sound betraying the presence of the apes. For ten minutes she wrestles with and giggles at the strange, playful foot with a mind of its own.

The mother is at ease to see her daughter happy, but she feels unwell. She has diarrhea. She spots a familiar bush, the tummy medicine tree. So she stirs to reach it, pulls down the growing tips, and spends an unpleasant couple of minutes chewing the green pith. The juice is vilely bitter, not something she would ordinarily even touch, but now she forces herself to swallow the medicine. She'll be better soon.

"Rhhooff!" Suddenly the peaceful moment is shattered by a hunting bark from a hundred yards away. Everyone lurches awake, up and running to see what's going on. Three males have cornered an unwary group of red colobus monkeys in a tall tree with only one good escape route. One of the hunters climbs high toward the terrified colobus group and rushes at them, pretending to come straight on but stopping short and shaking a branch at them, hoping to scare them into trying the escape route. One monkey jumps wildly, lands far away in another tree, and manages to scamper off. But the next one tries the more obvious escape and is snatched by a waiting chimpanzee as soon as he lands. The remaining monkeys pause in terror on their high perch. Now a hunter approaches, grasps their five-inch-thick branch, and furiously shakes it, until one, two, and finally three colobus jump or are shaken off. They leap and scamper frantically. And so it goes. Some chimpanzees press the attack, others focus the quarry's retreat, others wait in ambush, and soon the hunter apes have killed four monkeys.

A big male sits with meat in hand. Three other chimpanzees cluster around him, reaching eagerly with outstretched hands, watching anxiously for any sign of favor. One of the supplicants is his ally, his friend in the community's ongoing male status contests. So the meat owner tears a piece off his prize and drops it into his friend's hand. Encouraged by these signs of generosity, a female supplicant turns and invites the meat owner to mate. He does so, at the same time holding his valued property high to prevent a greedy hand from taking any. Then, after settling back, he rewards the willing female with a chew.

Another male, lacking meat to entice the attractive female, courts in his own special way from a few yards back. He picks a leaf and pulls at it, tearing the blade. She hears and sees him, and she understands the signal. So she goes to the sign-making male and mates with him, too.

Wild chimpanzees in the dappled forest teaching and learning, playing, communicating with invented signals, doctoring themselves, using tools to enrich their food supply. These scenes conjure classic visions of a peace in nature, an Eden of prehistory. Here is the brightly lit side of the picture, the angle we can all enjoy, and for more than a decade after Jane Goodall launched her study, it was the only side we knew. Like a rich fantasy by Jean Jacques Rousseau or a brilliantly colored canvas by Paul Gauguin, our first real view of chimpanzees was untroubled by any indications of serious social conflict. The apes seemed to wander without boundaries, with no fear of strangers. Sex was public, promiscuous, and unprovocative. There was little fighting over food. Science writer Robert Ardrey captured the mood of how we felt about them. Chimpanzees showed us, he wrote in 1966, an "arcadian existence of primal innocence." They stood for an idyllic past "which we once believed was the paradise that man had somehow lost."[6]

But then came Godi's death. It's true that for the most part chimpanzees lead very peaceful lives, but the attack on Godi raised the sudden possibility that chimpanzees had a dark side yet to be understood. Was this violence an aberration or the norm? Now, two decades later, we know the answer.

Jane Goodall was the first to watch chimpanzees at close quarters in the wild. She established her camp in a part of Tanzania's Gombe National Park known locally as Kasekela and began giving bananas to the chimpanzees in order to keep them near the camp. By 1966 she had identified fifteen different females and seventeen males in the area, as well as the young traveling with their mothers. All of these Kasekela chimpanzees interacted peaceably, and so Goodall thought of them as comprising a single community. But within that large community, so it slowly became apparent, two subgroups existed or were developing. Most of the individuals who showed up at the banana provisioning site came from all around, randomly; but a few tended to come from a particular direction, the south, and the southern chimpanzees usually stayed close to each other.

I arrived at Gombe in 1970, just as the north-south community division was beginning to be recognized. A zoology graduate student, I was supposed to be studying the relation between behavior and the food supply, but naturally I was drawn to the unfolding drama of the rival subgroups.

By 1971 the signs of a rift had become more obvious. During that year the eight adult males in the northern subgroup (Evered, Faben, Figan, Hugo, Humphrey, Jomeo, Mike, and Satan) and the seven from the south (Charlie, Dé, Godi, Goliath, Hugh, Sniff, and Willy Wally) met less and less.[7] When members of the two groups did encounter each other, there was obvious tension, particularly when Humphrey and Charlie were both present. These two were the dominant males of their subgroups and neither was willing to be friendly to the other, so any meetings involving them were launched by noisy, furious charges, followed by separated clusters of males grooming each other on opposite sides of the provisioning area.

By 1972 the only males nervy or oblivious enough to cross between the subgroups were Hugo and Goliath, the two oldest males, lifelong friends now weary with age. And by 1973 even this minimal contact had stopped.

So now there were two communities: the original one, Kasekela, and the southern breakoff, Kahama. It was a pity to see old friendships founder but, from the researchers' point of view, the split was interesting because now there were two communities unafraid of humans. For the first time we could watch interactions from both sides of a border.

Alongside other students and a growing corps of Tanzanian field assistants, I followed chimpanzees from dawn to dusk whenever opportunity and energy allowed. The long days took me to the far valleys of both communities' areas. I noted how in four-day cycles, parties composed of perhaps half a dozen males, sometimes with one or two females, would move tightly in patrols along the edges of their range in every direction.* I came to see where they tended to turn back toward the core of their range and how their behavior changed as they reached the boundaries. I learned where they were likely to stop and listen, and many times I heard them exchange raucous calls with males from the neighboring community. I saw how eager they were to embrace and grab each other in reassurance when they heard the exciting, alarming call of neighbors. I noted how, after listening to check that the other party was smaller than their own, they would rush forward to chase them from half a mile away; and sometimes they caught a neighbor and attacked him. Sometimes they made a mistake and charged toward a party that, despite sounding small at first, proved large, a situation that led to an immediate, confused, and hilarious retreat by the invaders

---

* *Party* refers to temporary groups, formed for a few minutes or hours, by chimpanzees from a single community. This is distinct from *troop*, which is a permanent group. Chimpanzees do not form troops, whereas many monkeys do.

back into the heart of their own land. I saw this latter event twice: the sudden conversion of a team of confident warriors into silent, scattered, nervous homebodies.

Comparing notes with my colleagues from the hot, exciting days of that year of discovery, I came to share with them a new view of male chimpanzees as defenders of a group territory, a gang committed to the ethnic purity of their own set.* By the fourteenth of August 1973, when a party of Kahama males led me to the freshly killed body of an unfamiliar adult female, we had fully accepted that these apes were ferocious defenders of community territory. Punctures on the victim's back indicated bites, her stretched body and tortured grip spoke of being dragged against her will, and a twisted final posture echoed the violence of her death.

Defense of territory is widespread among many species, but the Kasekela chimpanzees were doing more than defending. They didn't wait to be alerted to the presence of intruders. Sometimes they moved right through border zones and penetrated half a mile or more into neighboring land. They did no feeding on these ventures. And three times I saw them attack lone neighbors. So they seemed to be looking for encounters in the neighboring range. These expeditions were different from mere defense, or even border patrols. These were raids.

A raid could begin deep in the home area, with several small parties and individuals of the community calling to each other. Sometimes the most dominant male — the alpha male — charged between the small parties, dragging branches, clearly excited. Others would watch and soon catch his mood. After a few minutes they would join him. The alpha male would only have to check back over his shoulder a few times. The group would move briskly.

---

* *Territory* is a range that is forcibly occupied, i.e., defended from trespassers of the same (or sometimes other) species. *Range*, on the other hand, is a piece of land that is occupied whether or not force is employed.

Imagine it, then. The larger group, all or nearly all adult males, settles into a journey, stopping now and again to listen and look and rest. After a twenty-minute climb they reach a ridge, a border zone where they can either look back into their own valley or onward into the neighbors' range. They rest. Several climb trees. All are silent. All face the neighboring range. To the west, Lake Tanganyika sparkles through leafless trees.

After ten minutes they go on, more slowly now, cautious, newly alert even to ordinary sounds like snapping sticks. They leave the familiar range behind. They pause, listen. No calls from the neighbors. *Are the neighbors somewhere around, merely silent as we are? Or are they away in this dry season, eating the parinari fruits in the high valleys?* The party presses on.

Deep in the neighboring range now, they rest just before the top of a hill. Suddenly we hear footsteps on crisp dry leaves a few yards away, but the walker is hidden by the ridge and can't be identified. The raiding party freezes. The footsteps stop. Our alpha male is rigid, staring where he last heard the footsteps. The walker must be a few meters beyond, out of sight. *Resting? Aware of us? Is it a chimpanzee? If so, it has to be an enemy. There's danger here, because where there's one ape, there could be others. But what if they're alone? Or what if it's only a lone mother and infant?*

After six minutes of waiting for the walker to appear, the tension is too much for the alpha male. But he can't walk forward without crunching in the leaves, giving his position away just as the walker did. So he reaches forward and quietly grasps a sapling six inches above the ground. Then another with his other hand. Then another with a foot. He reaches a stump with his other foot. And so he steals forward, silent above the leaf litter until he sees who it is. *It's a baboon!* The tension drops, and he sits down without bothering to be quiet. *It's not an enemy. It's only a baboon....*

That was the sort of excursion that made chimpanzee territoriality seem more than defensive. The deep thrusts into neighboring lands weren't mere reactions, and they weren't in search of food: The raiders passed up chances to feed on the way and often fed only on their return. These raids were beginning to help us make sense of other problems. They explained why, whenever food supplies allowed, chimpanzees preferred staying together. There was power and safety in numbers. But why were these males raiding in the first place? That wasn't so clear.

Then came the killing of Godi. And seven weeks later, a second attack took place. Once again the victim was an isolated Kahama male — Dé was his name — and the attackers were a gang of four from Kasekela: three adult males and one adult female. One adult male, an adolescent male, and a younger female from Kahama watched from a short distance; clearly distressed, they were threatened away from time to time by one or more of the aggressors.

Upon sighting Dé, the Kasekela group rushed forward in obvious excitement, screaming, barking, and hooting, and surrounded their quarry. While the female of the raiding party, Gigi, screamed threateningly, the three Kasekela males closed in. Dé was helpless. According to the human observers, "He soon stopped struggling and sat hunched over, uttering squeaks." But at last he attempted to escape by climbing a tree, then leaping into another tree. Assaulted there, he fled onto a branch that broke under his weight and left him dangling low. From the ground, one of the Kasekela males was able to grab him by the leg and pull him down, and all three males, screaming, continued their assault. Finally Gigi joined in so that now all four were striking and stomping on the isolated male. They dragged him along the ground, biting and tearing the skin from his legs, and ceased their attack only after twenty minutes — having meanwhile driven away the two other Kahama males and, with threats, forced the young Kahama female to join their party. Dé was observed two months later, crippled, still severely wounded, and then never seen again. Missing, presumed dead.

One year later, a gang from Kasekela found their third victim. This time the target was Goliath, now well past his prime, with a bald head, very worn teeth, protruding ribs and spine. He may have been well into his fifties. It was many years since he had last competed for dominance. He had been a well-integrated member of the Kasekela community only five years before, and now (though he had since joined the Kahama group) he was little threat to anyone. But none of that mattered to the aggressors.

It began as a border patrol. At one point they sat still on a ridge, staring down into Kahama Valley for more than three-quarters of an hour, until they spotted Goliath, apparently hiding only twenty-five meters away. The raiders rushed madly down the slope to their target. While Goliath screamed and the patrol hooted and displayed, he was held and beaten and kicked and lifted and dropped and bitten and jumped on. At first he tried to protect his head, but soon he gave up and lay stretched out and still. His aggressors showed their excitement in a continuous barrage of hooting and drumming and charging and branch-waving and screaming. They kept up the attack for eighteen minutes, then turned for home, still energized, running and screaming and banging on tree-root buttresses. Bleeding freely from his head, gashed on his back, Goliath tried to sit up but fell back shivering. He too was never seen again.

So it went. One by one the six adult males of the Kahama community disappeared, until by the middle of 1977 an adolescent named Sniff, around seventeen years old, was the lone defender. Sniff, who as a youngster in the 1960s had played with the Kasekela males, was caught late on November 11. Six Kasekela males screamed and barked in excitement as they hit, grabbed, and bit their victim viciously — wounding him in the mouth, forehead, nose, and back, and breaking one leg. Goblin struck the victim repeatedly in the nose. Sherry, an adolescent just a year or two younger than Sniff, punched him. Satan grabbed Sniff by the neck and drank the blood streaming down his face. Then Satan was joined by Sherry, and the two screaming males pulled young Sniff down a hill. Sniff was seen one day later, crippled, almost unable to move. After that he was not seen and was presumed dead.

Three adult females, Madam Bee, Mandy, and Wanda, at one time had belonged to the Kahama group, along with their off-spring. But Mandy and Wanda eventually disappeared, as did their young, while Madam Bee and her two daughters, Little Bee and Honey Bee, were beaten by Kasekela males several times. Then in September 1975, four adult males charged the old fe-male, dragging, slapping, stomping on her, picking her up and hurling her to the ground, pounding her until she collapsed and lay inert. She managed to crawl away that day, only to die five days later. The assault on Madam Bee, incidentally, was watched by the adolescent Goblin and four Kasekela females, including Little Bee, who had become associated with Kasekela by then. Four months after Madam Bee was killed, her younger daughter, Honey Bee, also transferred to Kasekela.

By the end of 1977 Kahama was no more.

Horrifying as these events were, the most difficult aspect to accept was not the physical unpleasantness but the fact that the attackers knew their victims so well. They had been close com-panions before the community split.

It was hard for the researchers to reconcile these episodes with the opposite but equally accurate observations of adult males sharing friendship and generosity and fun: lolling against each other on sleepy afternoons, laughing together in childish play, romping around a tree trunk while batting at each other's feet, offering a handful of prized meat, making up after a squab-ble, grooming for long hours, staying with a sick friend. The new contrary episodes of violence bespoke huge emotions normally hidden, social attitudes that could switch with extraordinary and repulsive ease. We all found ourselves surprised, fascinated, and angry as the number of cases mounted. How could they kill their former friends like that?

Jane Goodall's early decision to provision the Gombe chim-panzees with bananas allowed observations that would other-wise have been very difficult to get. After evidence of chim-panzee violence appeared, though, some people suggested that aggression in Gombe was all the result of provisioning these wild apes with bananas: providing too dense a food source and thereby intensifying competition, promoting frustration, and ultimately bringing about unnatural behavior.[8]

However, even at Gombe, as it turned out, researchers soon saw aggression practiced by chimpanzees who had never been provisioned with bananas. After they had destroyed the Kahama community, the Kasekela chimpanzees expanded their territory into the Kahama heartland. At the edges of this newly expanded territory they met strangers coming up from a community in the south, from Kalande. Attacks ensued, and this time the Kasekela chimpanzees were the victims. Yet the aggressors had never visited Goodall's bananas — until one shocking day in 1982 when a raiding party from Kalande reached her camp. Some of the Kalande raids may have been lethal. Humphrey died near the border in 1981, his body found but his death unseen. Two infants died.

And elsewhere in Africa?

One hundred thirty kilometers south of Gombe, Toshisada Nishida has been studying chimpanzees in Tanzania's Mahale Mountains National Park since 1965, in the only chimpanzee research project other than Goodall's to have lasted for more than twenty years. And just as at Gombe, Nishida's team has seen border patrols, violent charges toward strangers, and furious clashes between male parties from neighboring communities. Once, in 1974, a male from one community was caught by three from another. He was held down, bitten, and stomped on, but he escaped. Is the violence at Mahale any less severe than at Gombe? Nishida doesn't think so. From 1969 to 1982, seven males of one community disappeared one by one, until the community was extinguished. Nishida and his team think that some, maybe most, of those who disappeared were killed by neighbors.[9]

On the other side of the continent things look much the same. In West Africa the first hint of intercommunity violence came in 1977, within Senegal's Niokola-Koba National Park, when conservationist Stella Brewer brought a group of ex-captive chimps into the forest with hopes of reintroducing them to a wild existence. But repeated attacks by native chimpanzees, including a terrifying nighttime raid of the camp by a gang of four adults, finally forced Brewer to shut her experiment down.[10]

Only a few hundred kilometers from Niokola-Koba, within the spectacularly rich Taï Forest of Ivory Coast, West Africa, Swiss scientists Christophe and Hedwige Boesch have been studying wild chimpanzees since 1979. Among the Taï chimps, territorial fights between neighboring communities were recorded once a month, on average; and the Boesches believe that violent aggression among the chimpanzees here is as important as it is in Gombe.[11] When an epidemic of Ebola virus reduced the number of adult males in the study community to two, Christophe Boesch feared that it would be taken over by a stronger neighboring group.[12]

In 1987 I joined Gilbert Isabirye-Basuta in his study of chimpanzees in the Kibale Forest, Western Uganda. One of the chimpanzees I came to know well, Ruwenzori, was by 1991 about fifteen years old, still the smallest and probably youngest of a clique of five teenaged males. In the second week of August, Ruwenzori was killed. No humans saw the big fight. We know something about it, however, because for days before he went missing, our males had been traveling together near the border, exchanging calls with the males from the Wantabu community to the south, evidently afraid of meeting them. Four days after he was last seen, our team found his disintegrating body hunched at the bottom of a little slope. The trampled vegetation bore witness to a struggle that started upslope and careened downward, sometimes sideways, for fifteen meters or more. Ruwenzori's body was bitten and bruised and torn. He died healthy, with a full stomach, on the edge of adulthood, on the edge of his range.

Kibale is providing the latest evidence that lethal violence, clearly witnessed in Gombe and strongly suspected in Mahale and Taï, is characteristic of chimpanzees across Africa. It looks like part of a species-wide pattern. In 1988 another apparently healthy chimpanzee died in the same border zone as Ruwenzori. At the time it seemed odd. We didn't know then where the border was. It seems less odd now. And three years after Ruwenzori's death, from only a couple of hundred meters away, we saw four Wantabu males stalk and charge a small Kanyawara party; but this time they caught no one. And then in 1994, one day after Kibale workers witnessed a violent attack on a male, tourists found the dead body of a prime male, probably the same victim. These Kibale attacks and killings have occurred in a forest where no artificial provisioning has taken place.

From the four research sites in the wild where chimpanzees live with neighboring groups, in work comprising altogether some one hundred years of organized field study, scientists have so far witnessed the extinction of two entire ape communities, certainly one and most likely both at the hands of their ape neighbors. In all four places the pattern appears to be the same. The male violence that surrounds and threatens chimpanzee communities is so extreme that to be in the wrong place at the wrong time from the wrong group means death.

The killer ape has long been part of our popular culture: Tarzan had to escape from the bad apes, and King Kong was a murderous gorilla-like monster.[13] But before the Kahama observations, few biologists took the idea seriously. The reason was simple. There was so little evidence of animals killing members of their own species that biologists used to think animals killed each other only when something went wrong — an accident, perhaps, or unnatural crowding in zoos. The idea fit with the theories of animal behavior then preeminent, theories that saw animal behavior as designed by evolution for mutual good. Darwinian natural selection was a filter supposed to eliminate murderous violence. Killer apes, like killers in any animal species, were merely a novelist's fantasy to most scientists before the 1970s.

And so the behavior of people seemed very, very different

from that of other animals. Killing, of course, is a typical result of human war, so one had to presume that humans somehow broke the rules of nature. Still, war must have come from somewhere. It could have come, for example, from the evolution of brains that happened to be smart enough to think of using tools as weapons, as Konrad Lorenz argued in his famous book, *On Aggression*, published in 1963.

However it may have originated, more generally war was seen as one of the defining marks of humanity: To fight wars meant to be human and apart from nature.[14] This larger presumption was true even of nonscientific theories, such as the biblical concept of an original sin taking humans out of Eden, or the notion that warfare was an idea implanted by aliens, as Arthur C. Clarke imagined in *2001: A Space Odyssey*. In science, in religion, in fiction, violence and humanity were twinned.

The Kahama killings were therefore both a shock and a stimulus to thought. They undermined the explanations for extreme violence in terms of uniquely human attributes, such as culture, brainpower, or the punishment of an angry god. They made credible the idea that our warring tendencies go back into our prehuman past.[15] They made us a little less special.

And yet science has still not grappled closely with the ultimate questions raised by the Kahama killings: Where does human violence come from, and why? Of course, there have been great advances in the way we think about these things. Most importantly, in the 1970s, the same decade as the Kahama killings, a new evolutionary theory emerged, the selfish-gene theory of natural selection, variously called inclusive fitness theory, sociobiology, or more broadly, behavioral ecology. Sweeping through the halls of academe, it revolutionized Darwinian thinking by its insistence that the ultimate explanation of any individual's behavior considers only how the behavior tends to maximize genetic success: to pass that individual's genes into subsequent generations. The new theory, elegantly popularized in Richard Dawkins's *The Selfish Gene*, is now the conventional wisdom in biological science because it explains animal behavior so well. It accounts easily for selfishness, even killing. And it has come to be applied with increasing confidence to human behavior, though the debate is still hot and unsettled. In any case, the general principle that behavior evolves to serve selfish ends has been widely accepted; and the idea that humans might have been favored by natural selection to hate and to kill their enemies has become entirely, if tragically, reasonable.

Those are the general principles, and yet the specifics are lacking. Most animals are nowhere near as violent as humans, so why did such intensely violent behavior evolve particularly in the human line? Why kill the enemy, rather than simply drive him away? Why rape? Why torture and mutilate? Why do we see these patterns both in ourselves and chimpanzees? Those sorts of questions have barely been asked, much less addressed.

Because chimpanzees and humans are each other's closest relatives, such questions carry extraordinary implications, the more so because the study of early human ancestry, unfolding in a fervor as we approach the century's end, is bringing chimpanzees and humans even closer than we ever imagined. Three dramatic recent discoveries speak to the relationship between chimpanzees and humans, and all three point in the same direction: to a past, around 5 million years ago, when chimpanzee ancestors and human ancestors were indistinguishable.

First, fossils recently dug up in Ethiopia indicate that over 4.5 million years ago there walked across African lands a bipedal ancestor of humans with a head strikingly like a chimpanzee's.

Second, laboratories around the world have over the last decade demonstrated chimpanzees to be genetically closer to us than they are even to gorillas, despite the close physical resemblance between chimpanzees and gorillas.

And third, both in the field and in the laboratory, studies of chimpanzee behavior are producing numerous, increasingly clear parallels with human behavior. It's not just that these apes pat each other on the hand to show affection, or kiss each other, or embrace. Not just that they have menopause, develop lifelong friendships, and grieve for their dead babies by carrying them for days or weeks. Nor is it their ability to do sums like 5 plus 4,[16] or to communicate with hand signs. Nor their tool use, or collaboration, or bartering for sexual favors. Nor even that they hold long-term grudges, deliberately hide their feelings, or bring rivals together to force them to make peace.

No, for us the single most gripping set of facts about chimpanzee behavior is what we have already touched on: the nature of their society. The social world of chimpanzees is a set of individuals who share a communal range; males live forever in the groups where they are born, while females move to neighboring groups at adolescence; and the range is defended, and sometimes extended with aggressive and potentially lethal violence, by groups of males related in a genetically patrilineal kin group.

What makes this social world so extraordinary is comparison. Very few animals live in patrilineal, male-bonded communities wherein females routinely reduce the risks of inbreeding by moving to neighboring groups to mate. And only two animal species are known to do so with a system of intense, male-initiated territorial aggression, including lethal raiding into neighboring communities in search of vulnerable enemies to attack and kill. Out of four thousand mammals and ten million or more other animal species, this suite of behaviors is known only among chimpanzees and humans.[17]

Humans with male-bonded, patrilineal kin groups? Absolutely. *Male bonded* refers to males forming aggressive coalitions with each other in mutual support against others — Hatfields versus McCoys, Montagues versus Capulets, Palestinians versus Israelis, Americans versus Vietcong, Tutsis versus Hutus. Around the world, from the Balkans to the Yanomamö of Venezuela, from Pygmies of Central Africa to the T'ang Dynasty of China, from Australian aborigines to Hawaiian kingdoms, related men routinely fight in defense of their group. This is true even of the villages labeled by anthropologists as "matrilineal" and "matrilocal," where inheritance (from male to male) is figured out according to the mother's line, and where women stay in their natal villages to have children — such villages operate socially as subunits of a larger patrilineal whole. In short, the system of communities defended by related men is a human universal that crosses space and time, so established a pattern that even writers of science fiction rarely think to challenge it.[18]

When it comes to social relationships involving females, chimpanzees and humans are very different. That's unsurprising. Discoveries in animal behavior since the 1960s strongly suggest that animal societies are adapted to their environments in exquisitely detailed ways, and obviously the environments of chimpanzees and humans are a study in contrast. But this just emphasizes our puzzle. Why should male chimpanzees and humans show such similar patterns?

Is it chance? Maybe our human ancestors lived in societies utterly unlike those of chimpanzees. Peaceful matriarchies, for example, somewhat like some of our distant monkey relatives. And then, by a remarkable quirk of evolutionary coincidence, at some time in prehistory human and chimpanzee social behaviors converged on their similar systems for different, unrelated reasons.

Or do they both depend on some other characteristic, like intelligence? Once brains reach a certain level of sophistication, is there some mysterious logic pushing a species toward male coalitionary violence? Perhaps, for instance, only chimpanzees and humans have enough brainpower to realize the advantages of removing the opposition.

Or is there a long-term evolutionary inertia? Perhaps humans have retained an old chimpanzee pattern which, though it was once adaptive, has now acquired a stability and life of its own, resistant even to new environments where other forms of society would be better.

Or are the similarities there, as we believe, because in spite of first appearances, similar evolutionary forces continue to be at work in chimpanzee and human lineages, maintaining and refining a system of intergroup hostility and personal violence that has existed since even before the ancestors of chimpanzees and humans mated for the last time in a drying forest of eastern Africa around 5 million years ago? If so, one must ask, what forces are they? What bred male bonding and lethal raiding in our forebears and keeps it now in chimpanzees and humans? What marks have those ancient evolutionary forces forged onto our twentieth-century psyches? And what do they say about our hopes and fears for the future?

These problems prowl at the heart of this book, and they are gripping enough. But they are made all the more curious by one strange, wonderful discovery of the last two decades. We have seen that chimpanzees and humans share, with each other but with no other species, a uniquely violent pattern of lethal inter-group aggression visited by males on neighboring communities, and we know that one possible explanation is inertia. As we will show in later chapters, the same applies to other patterns of violence, such as rape and battering. But one final fact destroys the theory that chimpanzees and humans share this appalling legacy merely by virtue of having shared a common ancestor who once behaved in the same unpleasant way. We know that inertia fails to explain similarities because chimpanzees have a sister species: the bonobo, or pygmy chimpanzee. Chimpanzees and bonobos both evolved from the same ancestor that gave rise to humans, and yet the bonobo is one of the most peaceful, unaggressive species of mammals living on the earth today.

Bonobos are critical to the vision we will develop in this book, and they are fascinating especially because of their remarkable females, who are in several ways more humanlike than female chimpanzees. Bonobos present an extraordinary counterpoint to chimpanzees, and they offer a vision of animals unlike any that we have been familiar with in the past. They have evolved ways to reduce violence that permeate their entire society. More clearly than any mere theory could, they show us that the logic linking chimpanzees and humans in an evolutionary dance of violence is not inexorable.

Bonobos, however, appeared late in the evolutionary timetable — just as they appeared late in Western science, and as they will appear late in our book. To understand how bonobos have changed the system, we must first understand the system. Still, bear in mind as we explore the patterns shared by chimpanzees and humans that the dark side will eventually be lightened by a strange species that wasn't even known seventy years ago, and wasn't watched until twenty-two years ago.

For the moment, our immediate journey from past to present covers more familiar ground. Surely we all know what chimpanzees are. They're the species so like us that we ask them to test vaccines for us, check the safety of space flight, or pose with us for photographs when we want to make fun of ourselves. Our closest relatives.

But what does that mean? Just how close are they?

2

TIME MACHINE

ARISTOTLE WAS THE ONE great philosopher who was also a biologist, so it's appropriate that he should have started it all.

In the fourth century B.C. he dissected some Barbary macaques (a tailless monkey species) and noted a remarkably close correspondence between monkey and human anatomy. Thus began a line of inquiry, proceeding in an erratic fashion up to the present day, that has drawn a slowly tightening net on the relationship between humans and other primates, and ultimately led to a scientific focus on one red-haired ape, the orangutan from southeast Asia, and three black-haired apes, the gorilla, chimpanzee, and bonobo, each from Africa. Aristotle called his Barbary monkeys "apes." But the *real* apes, those four great ape species, were only discovered by the Western world two millennia after Aristotle. They are strikingly more similar to humans than Barbary macaques are. This unavoidable similarity has made people wonder whether the great apes are our direct relatives, whether our ancestors looked and behaved like them, and what they can tell us about where we came from.

Ever since Darwin, questions and concerns provoked by the

very existence of those apes have rumbled through the corridors of science. Some were addressed long ago, so that, for example, it has become commonplace to think of humans as evolved from ancestral apes, albeit long dead ones. But the picture has never been satisfying because it never told us much about where we came from. The apes were one group of species; the hominid line was another. The two groups appeared to have been separate for so many millions of years, perhaps 10 or 15, that the significance of our common origin was lost in deep time. And the similarities in behavior between humans and any particular modern ape species, such as chimpanzees or gorillas, didn't mean anything in particular. They were just interesting and vaguely suggestive, hinting at what might have been. I remember well the strange sense of frustration enveloping discussions of ape researchers until recently. We would exclaim at the similarities between chimpanzee and human, and then pause, and say, "How odd!" But we were uncertain what to say next. The great apes were our closest kin all right. But they seemed too distantly related to tell us anything specific about our beginnings or our evolutionary journey.

Such was the wisdom up to 1984.

Then the world was blind-sided by a radical claim from two Yale biologists, Charles Sibley and Jon Ahlquist. Their DNA analysis placed humans, they said, right *inside* the great ape group. This extraordinary idea, if true, would demolish the concept of apes as their own discrete group separated from humans by a wide chunk of biological or evolutionary space. If the apes are a natural group, which we assuredly recognize them to be, then Sibley and Ahlquist's claim would make us a fifth great ape! Or, more precisely, the third, with two on either side. So imagine the family portrait now: five pairs of apes, one female and one male of each, orangutans and gorillas on one side, chimpanzees and bonobos on the other, and we humans stand in the middle of the photograph, happily flanked by our cousins.

You can say, if you like, that this new arrangement is only a theoretical juggling, hardly more than academic wordplay.

Whatever label one uses, you say, doesn't change the fact that the other apes are hairy and big-mouthed and unutterably crude in comparison with *Homo sapiens*. But for a very specific reason the new concept of humans as existing within the great ape group shakes our sense of separation from the animal world. The new theory of relatedness sends us back to the fossils and the living apes to examine once again the whole jigsaw puzzle of similarities and differences. And this time, with the help of a new perspective from modern genetics, the exercise projects a clear, startling picture. Suddenly we find a new time for our separation from the other apes, far more recent than we previously thought. And with the help of that new chronology, and a return to look once again at the other great apes, we now find humans descended not from some long-dead ancestor linking us only in a remote and unrealistic way to our modern cousin apes. Instead, the search for our own ancient ancestry pulls out at last an image terribly familiar and dauntingly similar to something we know in the contemporary world: a modern, living, breathing chimpanzee.

That's the claim. When the data from fossils and genes and living apes are combined, they give us a way to look back, to imagine the past with some real clarity. And against all our recent intuition they produce a device that tells us what we were like 5 million years ago when we left our African rainforest home and started the journey to humanity. They give us a time machine.

Blasted by dry heat, living in dirty makeshift camps in hostile countryside, fossil hunters searching for truth and glory in the African deserts have been so successful that the record of human ancestry over the last 4.5 million years is now one of the best for any species. It is so good, in fact, that even creationists accept the basic facts. Everyone agrees that some fossils are older than others, and that in the older fossils (which evolutionists date at 1.5 to 4.4 million years) ape and human features are combined.[1]

The record has become simply too clear for any sensible person to dispute these points.

Take a species like *Australopithecus afarensis*, the famous "Lucy." Fossil remains from Lucy's kind are found in northeast Africa, dated at 3 to 3.8 million years. Theirs was only one species out of a successful group, the australopithecines or woodland apes, as we will call them, who emerged almost 5 million years ago and lasted for about 4 million years.* Their bodies were about the size of a modern chimpanzee body. They had ape-size brains, ape-size mouths, and probably ape-size guts. Their hands and shoulders and upper body show they climbed well and could probably dangle from a single hand like modern apes.[2] But in some ways they were not like modern apes. The structure of their feet, legs, and hips unmistakably demonstrates that they walked upright almost as well as we do. And their teeth were not apelike. Their chewing teeth in particular were much bigger than those of humans or apes, though they were humanlike in being covered with a thick layer of protective enamel, unlike the thinly enameled teeth of chimpanzees and gorillas. Ape upper bodies, brains, mouths, and bellies. Human legs, pelvis, and tooth enamel. Quite a mixture.

---

* There is no ordinary name for the australopithecines. *Australopithecus* means "southern ape," so called because the first member of the genus was found in South Africa (the Taung baby, *Australopithecus africanus*, named in 1925 by Raymond Dart). We call them woodland apes because the known fossil habitats of the australopithecines were always more open than the closed rainforests used by gorillas and bonobos and by most chimpanzees. It's possible that some australopithecines lived in rainforests, or that some of their habitats were like the savanna woodlands occupied by certain populations of chimpanzees today, such as those in Mt. Assirik in Senegal, Ugalla in Tanzania, or even Gombe in Tanzania. Woodland apes, therefore, should be thought of as occupying a broad range of habitats, many of which would have been mosaics where there were deciduous woodlands next to riverine forest strips or more open areas of bushland or grassland.

These apes, perhaps six or more species, lived in the African woodlands for a stretch of around 4 million years. Their natural history is, to a large extent, our own prehistory, extending toward the present even later than the time — around 2 million years ago — when brains increased in size for one species of woodland ape, converting them into creatures with the earliest glimmerings of humanity.

We want to learn what we can about these australopithecines' lives, and where they came from. Strangely, a good way to start is by looking at the living apes.

For those who live in proximity with the modern apes, such as the native peoples of Central Africa, the idea that we are closely related to them is easy to accept, and apes figure prominently in some African myths of human origins.[3] But Europeans found this concept more abstruse. During the nineteenth century natural history became a popular topic in England, theories of evolution gripped the public imagination, and explorers and traders brought captive monkeys and apes to the zoos and menageries of Europe in numbers far larger than ever before. The public and the naturalists were entranced. Even so, when Darwin published his book about human evolution in 1871, *The Descent of Man and Selection in Relation to Sex*, he wrote that the majority of naturalists still rejected the idea that humans were primates. Instead, he claimed, most writers classified humans in their own order, the Bimana (Two-handed), conveniently separate from the order Primates.[4] In Darwin's view, this special self-treatment was absurd: "If man had not been his own classifier, he would never have thought of founding a separate order for his own reception."[5]

Perhaps Darwin was exaggerating about his fellow naturalists. It's hard at first to see how they could have kept humans out of the Primates (an order including monkeys, apes, and lemurs, among others), because a very convincing set of data already showed the extraordinary similarities between humans and apes. Published in 1863 by Darwin's friend and evangelist for evolution, Thomas Huxley, this first point-by-point comparison of humans with an ape produced such clear results that it remains a quintessential argument. Huxley compared humans with gorillas. He might have chosen another ape, but gorillas had just been discovered and were much in the news. This largest of the ape species was identified by Western science only in 1847, when a missionary physician returned to America bearing a collection of skulls and some African hunters' tales. Within a short time, the Philadelphia Academy of Sciences had packed a New Orleans newspaper reporter, Paul du Chaillu, off to find and shoot the awesome beast, so that during the 1850s crates filled with fresh gorilla skins and skeletons were being cracked open in the scientific depots of England and America and creating a storm of sensation and debate. Du Chaillu's subsequent best-selling account of gorilla hunting, *Explorations and Adventures in Equatorial Africa* (1861), stirred its own furious little whirlwind. The gorilla was, as Huxley wrote in 1863, "so celebrated in prose and verse, that all must have heard of him, and have formed some conception of his appearance."[6]

Huxley showed that in every kind of anatomical comparison, humans and gorillas are more like each other than either is to any species of monkey, the closest reasonable alternative. He showed, for instance, that human embryos start off looking just like those of other mammals, and then get different from more and more species. When the embryo still has a long, curved tail and looks like some strange little cube-headed alien, the yolk sac that feeds it is long and thin in dogs but round in humans and primates. And even after differences from other primates begin to emerge, the embryos of gorillas remain strikingly similar to those of humans. "It is only quite in the later stages of development that the young human being presents marked differences from the young ape," Huxley declared.

He similarly compared the adult anatomy of humans, gorillas, and other primates. He looked at limbs and vertebrae, pelvis and skull, brains and teeth, hands and feet. And in every case he showed that the differences between gorilla and human were fewer than those distinguishing a gorilla from any of the monkeys. Darwin's defender got it right. Apes are a small group of big primates that have no tails, limited to the gibbons from southeast Asia (the lesser apes) and the four great apes. They are not monkeys and are not even particularly closely related to monkeys. Monkeys form a much larger and more remote collection of species: baboons, macaques, colobus monkeys, langurs, and so on. Humans and apes are closely related to each other, while monkeys belong in their own group, separated, as we now know, from the ape-human line around 25 million years ago.[7]

One might imagine Huxley's methodical analysis to have been totally convincing. But it was attacked with great ferocity.[8] Much of the conflict occurred over issues that needn't concern us. But one particular weakness in his thinking was impossible to avoid, the weakness that always arises when evolutionary relationships are deduced solely from comparative anatomy. Although Huxley's extended comparison between gorilla and human anatomy did indeed powerfully suggest common descent, there was always another possibility, however remote. Through natural selection, unrelated or distantly related species can always evolve toward physical similarity as a response to similar environmental pressures. And this process, called *convergence*, can never be ruled out as an explanation for the anatomical similarity of species.

Consider, for instance, the fact that humans and the four great ape species all have rudimentary tail bones. The most likely explanation for this shared characteristic is descent from a common ancestor. Yet it is still possible that humans and the four apes descended from separate lines, with their rudimentary tail bones a result of each group independently moving into a similar environmental niche, thus evolving separately to some similar environmental pressure that promotes the discontinuation of a tail. Separate adaptation with convergent results. Could all the similarities between humans and great apes reflect this sort of convergent adaptation rather than a common ancestry? It becomes vastly unlikely, given so many points of similarity, but a strict anatomical comparison would never produce the ultimate proof. Darwin himself speculated with great skill, suggesting at one point that human ancestors were closely related to the fruit-eating apes from Africa. But he also knew where science stopped and speculation began. The possibility of convergence remained a critical limitation to any nineteenth-century conclusions about human ancestry. Early evolutionists needed to examine characteristics that don't respond as predictably to environmental pressure as most anatomical features do, and in the nineteenth century no one knew how or where to find them.

As so often happens in science, the breakthrough came from an unexpected quarter. George Nuttall was a polymath who among other things founded journals on hygiene and parasitology,[9] and, in his spare time, made pioneering discoveries in immunology. Between 1901 and 1904 this Californian tick expert lecturing in bacteriology at Cambridge University solved the convergence problem. He did so by carrying out blood tests on six hundred species, which showed that species that look alike on the surface (bones and brain and muscle) also look alike below the surface: They have similar blood chemistry.

By today's standards his technique was crude, but the approach was highly original and the results utterly impressive. Over a few weeks Nuttall gave rabbits five or six injections of human blood. What happens when a rabbit is injected with human blood Nuttall already understood: The rabbit's blood responds as if the human blood is a foreign invader like a virus or bacterium and makes antibodies to it. Several days after the last injection of human blood into the rabbit, Nuttall took a sample of the rabbit's blood, allowed it to coagulate in a bottle, and then drained away the clear serum floating above the clotted blood. That clear rabbit serum, he already knew, would contain antibodies against human blood, so he called it anti-human serum. In the same way, Nuttall produced rabbit samples of anti-chimpanzee serum, anti-orangutan serum, anti-mouse serum, and so on.

Now he took each serum specimen and added fresh blood from another species. Serum sample plus blood reacted together, making a precipitate. But the strength of the reaction, how thick the precipitate, depended on which species the blood had come from. The closer the relationship between blood and antibody serum the stronger the reaction; the more distant the relationship the weaker the reaction. So when Nuttall added human blood to anti-human serum, the reaction was very strong. But mixing human blood with anti-mouse serum produced no reaction. The tick expert had created an index of evolutionary relationship.

What happened when Nuttall added human blood to anti-ape serum? It made a thicker precipitate than human blood and anti-monkey serum did. All his results pointed the same way. This first study in molecular evolution showed its power in its unambiguous result. It placed humans closer to apes than to monkeys.[10] Huxley's hypothesis was confirmed. Nuttall's index of evolutionary relationship based on blood protein showed that the anatomical similarities between humans and apes were due to common ancestry, not convergence.

Convergence is no longer a theoretical worry in dealing with molecules like blood proteins because when environmental pressures shape a species in one way or another, that shaping affects anatomy and physiology without reaching down to the molecular level in the same way. Witness the striking examples of convergence between distantly related species, such as the European placental mole and the Australian marsupial mole, or the Tasmanian wolf and the timber wolf, or dolphins and ichthyosaurs, all of which look remarkably like each other but molecularly are very different. When you find real similarity in molecular design, the probability of convergence is vanishingly small. It's like comparing an English version of the *Odyssey* to the Greek original. No one would think to ask whether the English version is a translation from Homer or an original poem by an English author that just happens to tell the same story with the same allusions and the same poetry and the same characters. When the similarity runs that deep and broad and complex, common ancestry can be the only explanation. It must be a translation from the Greek, not an English original.[11] Likewise, if humans and apes have similar blood chemistry, elementary probability theory says it can mean only one thing: common genetic origin.

Nuttall completed his groundbreaking work in 1904, and over the next several decades his conclusion that the great apes were the closest animals to humans was amply confirmed by other kinds of biochemical studies.[12] This work naturally prompted people to wonder whether any one of the ape species was closer than the others to humans. Every kind of possibility was put forward: that humans were closest to orangutans, or chimpanzees, or bonobos, or gorillas, or equidistant between two, such as gorillas and chimpanzees.[13] But the question couldn't be answered because no one could find enough differences in the molecules being studied. The ape group and humans were simply too close for such details to be resolved by the kind of data then available. A new kind of data source was needed, which many people thought could only be a continuous series of fossils connecting, like footprints in the snow, our modern selves back to our early ancestors and forward to the living great apes.[14]

But advances in biochemistry opened up more and more ways to document differences between and connections among species. Everyone already knew that genes are responsible for the biological differences between species, but no one had a way to measure the differences between genes directly. Then, in the early 1960s (around the time Jane Goodall was first raising her binoculars to look into the lives of wild chimpanzees in East Africa), laboratory scientists in Europe and America were discovering that genes determine the structure of proteins. Was it possible, then, that protein structure could help map genetic associations between species? The ideal analysis would have looked at the order of amino acids, the building blocks of proteins, but no method had been devised for doing that. However, an alternative probe was available. Because they possess their own small electrical charges, proteins placed on a specially prepared medium move under the influence of electricity. Their speed depends upon their own charge, size, and shape. Seizing on this as a way to differentiate proteins from related species, scientists found that the proteins of humans and the three African apes are structurally more similar to each other than to the proteins of orangutans. These big red apes from Borneo and Sumatra had often been thought of as just another great ape species, albeit the only one living outside Africa. Laboratory magic now broke apart the ape group. It put humans right alongside the African apes; and it left orangutans dangling by themselves a little farther out on a branch of the evolutionary tree.[15]

There remained one more level of discrimination: direct DNA analysis. Protein studies were certainly very useful; and proteins, as products of DNA, provided some fresh hints about the almost infinite intricacies wrapped inside the genetic material itself. But DNA varies more than proteins do, and therefore DNA offers more information than proteins can.

Up to this point, as a matter of fact, the biochemical studies had been useful in dismissing convergence, but the species relationships they revealed were not so surprising. Orangutans are patently different from the African apes, and a century before, Darwin had suggested that humans were linked evolutionarily to the African apes. As for the three African ape species — gorillas, chimpanzees, and bonobos — well, a little common sense combined with some elementary anatomy would indicate that they were tightly associated in their own little group. After all, the African apes have several special features in common. Unlike orangutans or humans, they walk on their knuckles. They have thinly enameled teeth. They are black-haired. Surely, so went the conventional thinking up to 1984, they were all each other's closest relatives, with humans as the special outsiders whose ancestors had peeled off first.

DNA is an extraordinary molecule with a very simple structure. It is huge, very long but very thin, made of two halves that fit tightly together, like a mile-long zipper. The teeth of the zipper are chemical units that bond with the partner opposite, the complementary zipper tooth. The zipper's teeth are chemical units called nucleotides, and they come in four and only four types: adenine and thymine, cytosine and guanine. Each type bonds only with one of the others. Adenine bonds only to thymine (and vice versa), and cytosine only to guanine (and vice versa). This means that each half of the zipper is utterly predictable from the other half. So all you have to do to analyze the similarity between the DNA of two species is take half of the zipper and read the nucleotides. More closely related species have a more similar list of nucleotides.

In 1984 no one could do that. Even now it's a slow and painful process, so that a single laboratory identifying nucleotide sequences will tackle only relatively short DNA segments (10,000 to 15,000 nucleotides long) — in other words, single genes, the units that code for single proteins. But the primitive state of nucleotide identification in 1984 didn't deter Charles Sibley and Jon Ahlquist. Rather than wait for the invention of more sophisticated laboratory equipment in order to identify genetic structure, they used the equivalent of Nuttall's technique, examining, by means of a technique they called DNA hybridization, how well genetic strands from different species fit together.

With already standardized chemical procedures, Sibley and Ahlquist extracted single strands of DNA from the blood of two different species and then allowed the two strands to bind, or "zip" together. They were experimentally creating hybrid DNA zippers that — depending on the relatedness of the two contributing species — would contain some level of imperfection in the zip. Then they applied heat. Gradually raising the temperature stressed the DNA zipper, slowly causing the strands to separate. Hybrids between closely related species formed stronger bonds, so it took a higher temperature to separate them. And this "melting temperature" thus provided a yardstick for how closely related species were.

Sibley was a professional ornithologist; Ahlquist, his assistant. Their first studies, therefore, compared birds. The relatedness among bird species had already been carefully worked out to the satisfaction of most ornithologists, and for birds, Sibley and Ahlquist's DNA hybridization worked beautifully. Their genetic data matched well with the evolutionary relationships already established by comparative anatomy.[16]

After this first demonstration with birds, the two researchers turned to the human-ape relationship. Two of their questions are particularly relevant to us now. The first didn't produce much surprise: *Which species is closer to humans: chimpanzees or gorillas?* Sibley and Ahlquist took strands of DNA, zipped them together to make human-chimpanzee and human-gorilla hybrids, turned up the heat, and then looked at the results. The answer was *chimpanzees.* Fair enough. Chimpanzees are closer to human size. They use tools more than gorillas do. And in general they look a bit more human. The answer was reasonable, though it could have gone either way.

The next question produced the shocker: *Which species is closer to chimpanzees: humans or gorillas?* Obviously, chimpanzees and gorillas look very much alike. And humans look very different from both. So, naturally, everyone expected that chimpanzees and gorillas would be each other's closest relatives. But Sibley and Ahlquist took the two strands of DNA, zipped, heated . . . and found that chimpanzees were more closely related to *humans* than they were to gorillas.[17]

Hardly anyone believed it. Critics pointed out problems with the analysis. So Sibley and Ahlquist, properly chastened, reanalyzed their own data. The results held.[18] New methods were called for.[19] The thermal stability of DNA hybrids was measured with a better technique. Yet the results were still almost identical to Sibley and Ahlquist's original work.

More convincingly, while the DNA hybridization technique was being refined, new methods allowing direct analysis of individual genes by listing the sequence of nucleotides extended and reaffirmed the Sibley-Ahlquist results. Even as we write, new and ever more precise data on the genetic connections between humans and apes are being published regularly. The entire mitochondrial genome has been sequenced, and several nuclear genes. Virtually all studies point the same way, and the few exceptions are easily explained.[20] To all but the most hardened defenders of the old view, the battle is over. The geneticists' ability to read the genetic code has turned upside down the old picture of human-ape relationships. The new picture is this: The two species most closely related are chimpanzees and bonobos. Next to this pair is humans. And then come gorillas, followed at a distance by orangutans.

With the overwhelming DNA evidence now before us, where can a skeptic turn? For anyone who still doesn't like the idea that we humans are more closely related to chimpanzees than gorillas are, what can be done? In the last century, the ultimate skeptical response to the discovery of fossils was to decide that God must have put them into rocks as an aesthetic or philosophical experiment, to pretend that the earth has a history in the same way that He gave Adam a navel to pretend he had been born of woman. For some creationist skeptics of contemporary times, on the other hand, an evil Satan laid down all those fossils in order to tempt us into accepting evolutionary theory.[21] Likewise, one could argue that the molecular marks of a deep relatedness between humans and apes serve either a divine or a devilish plan. For most of us, however, the idea of a deceitful power working at that level strains credulity too severely. Omnipotent the Creator may be, but not weird.

The Sibley-Ahlquist result makes a time machine possible.

First and most obviously, genetic relationships echo evolutionary history. If gorillas are most distantly related to the rest of the African ape line, for example, then we know that ancestral gorillas split off from this line first. Since humans are closer to chimpanzees than gorillas are, we know that humans diverged next, after the gorillas did. Finally, since chimpanzees and bonobos are most closely related, we deduce that the ancestors of chimpanzees and bonobos separated most recently of all.

Before Sibley and Ahlquist's work, the common ancestor of humans and living apes was supposed to have lived 10 to 15 million years ago, maybe more. That would have meant our common ancestor with the modern apes was lost in a stratum of time where the relevant fossils are few and hard to interpret. But now, according to the new data, our common ancestor with the modern apes lived recently enough to approach the era of the woodland apes; and for the woodland apes our fossil data are rich and relatively coherent.

We can estimate the time of our common ancestor more precisely by comparing genetic similarity with species whose dates are already well established from reliable portions of the fossil record. The standard baseline is the divergence of orangutans from African apes, now reliably dated from good fossils at 13 million years but conceivably somewhere in the 10- to 16-million-year range. Using that date as a benchmark, genetic information from more than 11,000 individually identified nucleotides in the mitochondrial genome indicates that the common ancestor of chimpanzees and humans lived some 4.9 million years ago. On either side of this figure fall estimates produced from other sources: using nuclear DNA sequences, 4.6 million years ago; using DNA hybridization, somewhere in the 5 millions.[22] Imprecision remains, but the real date of divergence — that amazing waking moment from the dream of time when two individual apes, one from a group destined to evolve into humans, the other from a group destined to become modern chimpanzees, shared a final moment of mutual recognition and ambled off in separate directions — lies between 4.5 and 6 million years ago. We call it 5 million.

Only 5 million years! The oldest known fossils of woodland apes go back to 4.4 million years. So these woodland apes were living only a short time — perhaps a few hundred thousand years — after our common ancestor with the rainforest apes. If we can now imagine what our rainforest ape ancestor looked like, then we may picture our family tree almost all the way from rainforest ape to woodland ape to early human.

To picture our common ancestor with the rainforest apes living 5 million years ago, you might think we should simply blend chimpanzees and humans into a single image, a mixture of the two species. You'd be wrong.

The problem with that approach is that it presumes comparable rates of change. But humans, in an evolutionary sense, are a radical species; we have evolved particularly fast — not as fast as polar bears, say, which evolved from grizzly bears around 20,000 years ago, or as dwarf deer, which when isolated on islands lose their mainland body size too fast for paleontologists to count the years — enormously faster than most species. *Homo sapiens* is only about 150,000 to 230,000 years old, and we have proceeded through enough shape-shifting to define at least four other, prehuman species in our background before getting back to the rainforest apes.[23] We've changed so much so quickly that our present doesn't tell us very much about even our recent past.

The right way to consider what our ancestor was like is to focus on the rainforest ape. We know it was an ape. What kind of ape was it? At 10 million years ago its ancestor had spawned the gorilla line. Now, at 5 million years ago, it was dividing in two to produce both the chimpanzee and the human lines. Can we imagine what our ancestor looked like 5 million years ago?

Yes, we can, because gorillas and chimpanzees look so alike. Even the experts can have trouble telling the difference. When the American explorer and gorilla hunter Paul du Chaillu visited England in 1861, he presented to the British Museum the re-markable skull from, so he insisted, a previously unidentified ape species that Gabonese hunters called the kooloo-kamba. Du Chaillu described his specimen skull as taken from a male "smaller than the adult male gorilla, and stouter than the female gorilla." With a markedly high forehead and unusually large cranium, more closely resembling a human than any other ape, the kooloo-kamba was yet most distinguished by its cry: *koola-kooloo, koola-kooloo.*

Du Chaillu concluded that the kooloo-kamba was neither a chimpanzee nor a gorilla, but rather a "chimpanzee-like animal," apparently a new subspecies of chimpanzee; but some later commentators thought the skull big enough, and enough like a gorilla in other features, to suggest either a species intermediate between the gorilla and chimpanzee or, more simply, the hybrid offspring of a gorilla-chimpanzee mating. This debate occurred most intensively during the latter part of the nineteenth century, in part because several apes then in captivity seemed to evoke the same issue. Mafuca, a female ape acquired by the Dresden Zoo in 1874, for example, confused a number of experts, who had trouble deciding whether she was a large chimpanzee or a small gorilla. Others considered her a hybrid, while one prominent British anatomist, Sir Arthur Keith, pronounced Mafuca to be a living specimen of kooloo-kamba. An adult female ape named Johanna, owned by the Barnum and Bailey circus around the same time, likewise confounded the authorities by representing, in the words of one, "a variety of chimpanzee which approaches the Gorilla in so many points that it is evident the characters which separate the two African anthropoids are not so well marked as many suppose."[24]

Du Chaillu's skull turned out to be from a large chimpanzee, while other supposed kooloo-kambas were sometimes large chimpanzees and sometimes small gorillas. The extent of the debate now seems surprising; but the story has special significance for us here because it demonstrates how physically alike gorillas and chimpanzees actually appear. Modern gorillas and chimpanzees profoundly resemble each other — except in size.[25]

(In fact, DNA comparison shows the separation of these two species much more thoroughly than any anatomical or social examination can. DNA builds all sorts of internal structures, including brains, that we cannot ordinarily see; moreover, the genetic record includes huge amounts of "junk DNA" produced by random mutation, stored but functionally inactive.)[26] In appearance, then, unless we postulate a remarkable phase of divergence and then convergence, the rainforest ape line has changed very little since these two species split around 8 to 10 million years ago. That means that our own rainforest ape ancestor, peeling away from the same line at 5 million years, came out of the chimpanzee-gorilla mold. If our rainforest ape ancestor was big, it should have looked like a gorilla. If small, a chimpanzee.

Well, then, what size was the ancestral rainforest ape 5 million years ago? It's a guess, but the oldest woodland ape fossils, dated at 4.3 to 4.5 million years, fit a creature weighing essentially the same as modern chimpanzees.[27] So the best guess is that around 5 million years ago the rainforest ape ancestor that diverged to produce modern humans from one line and chimpanzees from another was probably the size of a chimpanzee.

Chimpanzee-size, and made in the gorilla-chimpanzee mold. So what did it look like? A chimpanzee. That is the startling prediction for the missing fossil, the ape of 5 million years ago that launched the transition from rainforest to woodland.

Evidence in support of this prediction has already appeared. In September 1994 the fossil remains were identified from a new species of woodland ape, *Ardipithecus ramidus* (or *Australopithecus ramidus* as it was first named, before more fossils were found in late 1994).[28] Older than Lucy by perhaps a million years, *A. ramidus* lived around 4.4 million years ago and looks more like a chimpanzee than Lucy in nearly every feature.[29] There has even been discussion whether *A. ramidus* should have been called *Pan ramidus* — a chimpanzee species! As the fossil hunters push further back into the past, up to 5 million years or a little more, we predict that their finds will look even more like a chimpanzee.

Eager to contemplate our past, Darwin was nevertheless strict to deter people from thinking that modern species could be treated as equivalent to our ancestors. "We must not fall into the error of supposing that the early progenitor of the whole Simian stock, including man, was identical with, or even closely resembled, any existing ape or monkey."[30] Darwin was right to be careful. Textbooks routinely caution against this naive error in thinking. And, in most cases, the warning is appropriate.

But occasionally the normal rule of dissimilarity between ancestors and living relatives can be broken. Evolutionary change occurs at different speeds in different species. Some are conservative, evolving little during a long period of time, surviving happily in a stable environmental niche. Others are radical evolvers, responding dramatically to dramatically shifting environmental pressures. Chimpanzees seem like a conservative species, one that has changed remarkably little over the last 8 to 10 million years, perhaps because the essential adaptation of these apes is to large tracts of tropical rainforest.[31] The particular tree species don't matter; all the equatorial forests produce abundant fruit, so all are habitable by apes. Such forests have provided a continuous home for fruit-eating primates throughout the era of modern mammals, ever since life on earth was decimated by a 10-kilometer-wide asteroid slamming into the Yucatán peninsula at the end of the Jurassic, 65 million years ago.[32]

So a warm, wet, fruit-rich forest has lain in the heart of Africa throughout the era of the apes. In response to climatic vagaries, it has sometimes expanded and sometimes retreated, and perhaps the canopy has become thinner and allowed more foliage in the underlayer, but it has always survived in a form that apes can use. The modern great apes — orangutans, gorillas, chimpanzees, and bonobos — split away from each other during the last 15 million years. They are, we suggest, all evolutionary conservatives because they live in conservative habitats. And when we turn to chimpanzees, we may be gazing upon as conservative an ape as any, a species that has roamed and exploited African tropical forests with very little change for 5 to 10 million years.

Such is the argument for thinking that to be with modern chimpanzees in an African rainforest is to climb into a time machine. Stepping into the dappled world of these extraordinary apes we move back to glimpse our origins. It's not a perfect picture, but it's amazingly good. The essential forest structure is the same as it was then. And the apes enduring within it have remained likewise much the same: knuckle-walking, arm-hanging, large-brained, heavily built, black-haired, big-mouthed, and fruit-eating.

How were our chimpanzee-like ancestors behaving 5 million years ago? The Sibley-Ahlquist result makes our shared characteristics more likely to come from common descent than from convergence because they explain why our behavior should be more similar to chimpanzees than to gorillas. We have already described how humans and chimpanzees, but no other species, share a unique combination of social characteristics: male-bonded communities and male-driven lethal intergroup raiding. If two closely related frogs share a unique set of behaviors, the coincidence of kinship and behavior would make us think the behaviors have a common origin. Apply that logic to ourselves, and we would say that our own male-bonded, war-shaped societies have evolved for whatever reasons they did in chimpanzees during the time of our common ancestor or earlier. Five million years ago, this suggests, there were killer apes.

But it doesn't prove it, and it does not mean that we have explained why shared characteristics occur now or why our ancestors should have had them. Those questions we must answer head on, by asking what the behaviors do for their perpetrators.

We'll address those issues. But first let's look at the gap between chimpanzees and humans in one more way. It's all very well to assert that chimpanzees and humans are close kin, but what happened in Africa 5 million years ago to split our ancestry so distinctively? What steps link and divide our two species more particularly? What pressed one small group of rainforest apes toward the woodland, and from there to start the human line? And what will that vision have to say about the way the species' violence was suppressed or maintained during our long sojourn as woodland apes?

3

ROOTS

L ET US IMAGINE OUR 5-million-year-old ancestors, then, as rainforest apes barely distinguishable from chimpanzees. Much as chimpanzees do now, they occupied the equatorial forests from the Atlantic across the continent into eastern Africa. Even then they were shouting with excitement and pleasurable anticipation as they arrived at fruit trees, grunting softly to each other as they lay in their nests waiting for sleep, waving leaves at each other to induce a chase, tickling their babies, hunting monkeys, and fighting over turf. Some groups probably collected ants with sticks, some washed fruits in pools, some toyed with leaves to show an interest in mating. For two hundred thousand generations, we can readily imagine, they continued inventing and forgetting and reinventing ways to make life a little easier. Over the generations dry seasons sometimes grew longer, sometimes shorter; the rainforest retreated or expanded; habits and traditions came and went; other species cycled through their lives; and the world turned, until chimpanzees washed up on the doorstep as our neighbors today.

But around the time we're considering, as the Miocene epoch was giving way to the Pliocene, one little chimpanzee-like community and its culture was jolted out of stasis. Probably our own Adam and Eve started life on the edge of the rainforest ape species' range, maybe in the region that today is called Ethiopia.[1] Their forebears would have emigrated during a wave of forest advance from the south, in a time of high humidity and global warming. But now the climate has turned. A long drought has settled on the continent. Away from the equator, the tall fruit-bearing trees no longer replace themselves and are outcompeted by dry-country species relying on winds, not animals, to spread their seeds, and so they give little food reward to fruit eaters. In the dampest areas, where river gullies nestle in protective gorges, a few islands of rainforest provide the last refuges for these apes now surrounded by a spreading ocean of hostile savanna woodland. Eventually, even those little nurturing streams dry up. The loss of rainforest means the loss of food. Breeding will soon cease, and the localized ape populations will succumb, go extinct. It's a common fate for pioneers like these who have spread from their equatorial base to higher latitudes. But pioneers sometimes succeed, and success is what happens to this single group of apes we're thinking about. They are lucky. Something allows their small band to survive and change.

How did at least one population of rainforest apes turn their dwindling refuge into a launching pad for adaptation to the woodland?

Bipedal travel is considered by most people the hallmark of the hominids — at least in the sense that if a fossil ape species was walking upright on two legs, it's called a hominid. Chimpanzees stand erect and even walk upright on occasion, as do all the other apes to one degree or another. Usually chimpanzees do it when forced by circumstance, when, for example, a hand is injured or their arms are full of fruits or they drag a branch in display. But they don't have the right skeletal and muscular structure to make this posture easy. A chimpanzee like Gombe's Faben, who lost the use of one arm to a poliolike disease, often walked upright, sometimes using his good hand as a third leg; the style was functional but nevertheless awkward enough to make him frequently lag behind his peers. For our rainforest ape ancestor to evolve from four-legged to two-legged would have required some critical new circumstance making uprightness so useful that, in spite of its likely awkwardness and discomfort during a long period of evolutionary transition, some individuals did it enough to acquire for themselves a steady advantage. Unfortunately, we have no good explanation yet for what that critical circumstance was, for why the new woodland apes adopted two-leggedness as their dominant stance. Or, if you like, there are too many explanations, with none yet clearly better than the rest. To free hands for carrying? To reach and pick berries higher up? To keep the sun off their backs? To gain height to see farther? To wade through swamps? There's something to be said for each idea.[2] Our favorite will emerge below.

Whatever its initial advantages, bipedalism was not only an identifying characteristic of the new apes, it also critically influenced their changing lifestyle. Once our apes evolved far enough anatomically to pass the stage when bipedalism was awkward, during a transition lasting for hundreds of generations or even longer, they could go much faster and farther than chimpanzees can — twice as far in a day, roughly.[3] The greater distances would have been useful for finding the most productive feeding areas in the drier, more open woodlands where these apes now lived. Not to mention escape from speedy predators, such as a monstrous carnivorous bear that shared their woodlands for half a million years in the early Pliocene.[4]

Even so, there is an issue more critical than how they traveled. Bipedalism probably wasn't essential to woodland life for our apes. After all, the woodlands support any number of other species that aren't upright walkers. What they ate in their new habitat must have been much more important. Changing locomotion style could have taken place over a long time, but changing food habits was required immediately. Food is the *sine qua non* of animal life. In addition to the obvious importance of food for daily survival, a species' food supply profoundly influences many of its other adaptations. Food explains not just tooth structure but overall population size, social group size, competition style, ranging patterns, and so on. Giant pandas have huge, flat teeth because they eat bamboo. Birds migrate to find the best food. Wildebeest live in great herds because their food is in huge patches. Big, fierce animals are rare because their food is scarce. . . . You get the idea. To understand what happened at this epochal moment in our own species' history, we need to imagine their menu.

Today chimpanzees live only in areas where there is enough damp rainforest to provide several square kilometers for each individual.[5] Such forests are often fringed or even surrounded by savanna and woodlands, and chimpanzees use these more open areas enthusiastically. Foods are occasionally plentiful there. A week-long bounty of immature seeds in an orchard of bean trees can attract long expeditions, leading sometimes to overnight stays. So woodland is not inherently hostile or useless for a rainforest ape. But it is still only an option, whereas rainforest is a must. Rainforest apes need rainforest foods, which means soft fruits and wet foliage. No woodland provides year-round foods for a chimpanzee.

To colonize the woodlands, the apes required a new food habit, an inclination and ability to eat some particular item that occurs there predictably. This food item must have some remarkable properties. It must be so rare in the forests that most rainforest apes have not already adapted to using it. In the woodlands, however, it must occur at sufficient density throughout the year to be a reliable food source. It must not be too heavily eaten by other species already living there. It must provide a secure enough food base to allow for the evolution of several species of woodland apes, and eventually for those three or four genera that would continue to use it. It must explain why all the hominid descendants of those ancient woodland pioneers would share, to varying degrees, an identifying pair of dental features: first, thickly enameled teeth in general, and second, especially massive, round-edged molars or cheek teeth. Without this new food, the emerging rainforest apes would have been, as modern chimpanzees still are, dependent on the ripe fruits and soft herbs that only well-watered rainforest provides sufficiently.

The food would have been rare in the rainforest, abundant in the savanna woodland, underexploited by other woodland animals, present even when other foods give out, and best processed by large teeth covered with a thick layer of enamel. What was it?

Various candidates have been promoted — seeds and nuts, for instance. Thick enamel would serve to protect teeth from fracturing when crushing nuts, as demonstrated by those seed-eating monkeys with thickly enameled teeth. Apes would be better than most species at opening nuts or bean pods, using both their strength and their invented tools. And many woodland trees produce fat-rich seeds or protein-rich beans. So a seed and nut specialization was possible, but we don't think it was the solution. Seeds and nuts seasonally disappear from the woodland, so this food source could not sustain apes the year round.[6] And there would be little new about this adaptation to explain why rainforest apes didn't and don't regularly move onto the woodlands. The ancient apes presumably already ate seeds and nuts, after all, much as chimpanzees do now in both rainforest and woodland. And why would relying on seeds and nuts promote evolutionary selection for teeth that are large? Other seed eaters don't have particularly large teeth, merely thickly enameled ones. And would seeds and nuts be common enough to carry the woodland apes through all the different types of habitat where their bones are found in the fossil record, including those with few trees? Probably not. No, seeds and nuts could not explain how woodland apes found enough food to survive.

Meat? Our common ancestor with chimpanzees presumably already ate meat. Modern chimpanzees love meat. But even though their woodland habitats were sometimes bursting with edible colobus monkeys, the emerging apes would surely never have caught enough to satisfy their nutritional needs, especially now that their bipedal walking reduced them to being slightly less effective climbers than before. Perhaps, though their brains were no bigger than those of chimpanzees, they devised better systems for catching antelopes or hares. But the teeth of carnivores are sharp, not broad and blunt like those of the woodland apes. If there was a big commitment to meat eating it surely came 3 million years later, when *Homo* evolved.[7]

What else? Grass seeds? Only there for a few months each year. Tree leaves? Few are edible. Bones? Unlikely to be enough. A little of everything? But why the particular tooth form? And so on. All classes of foods present unresolved problems except one: roots.

What if one population of apes during the drying period had already developed, while still living in the rainforest, a tradition of finding and eating carbohydrate-rich roots? Roots would always have been present beyond the rainforest, as they are today in huge densities wherever they have been measured in savanna woodlands, so they could provide great concentrations of food energy. They would be eaten by few other animals, hidden and often hard to retrieve then as they are today. Many, like those still consumed by human foragers, could be eaten raw. They would be present all year round, providing reserve food supplies during hard times. And broad, thickly enameled teeth make sense as an evolved adaptation for eating them. Roots could indeed have been the fallback food that carried our ancestors, and therefore our genes, from dwindling forest into the more open woodland and savanna.

The Ituri Forest, in northeastern Zaïre, is one of the largest blocks of rainforest in Africa and an intriguing place to think about human evolution. Dotting the expanse of rich green are occasional granite outcrops, called *kakbas* by the local inhabitants, that break through the canopy sometimes by a few meters, sometimes by much more. These *kakbas* offer an escape from the claustrophobia of trees. Sweating your way to the top, you find yourself in a little world within a world. Your eye stretches to a tree-rippled horizon dauntingly flat and far, its flatness occasionally made imperfect by another *kakba*. The still, dim, humid world of the rainforest floor has been left behind.

Here on the *kakba* it is dry and sometimes windy. In the cracks between the rocks you see grasses, and on the rocks you notice the droppings of buffalo and hyrax. Drought-adapted plants with swollen stems and leaves testify to the need to store water before it runs quickly off to join the sodden rainforest floor. The *kakba* is light and cool, a little island of dry savanna woodland in a sea of wet forest. It rewards the climber with breezes and some perspective over the rest of the rainforest. And it provides perspective on our problem: How could a rainforest ape colonize the woodland?

From the top of the *kakba* you will sometimes see smoke rising in the distance. Farmers live in the rainforest, and the smoke rises from patches they burn out in order to plant crops. But Pygmies also live in the Ituri, and they survive by using the rainforest in other ways. Pygmies are famous for their hunting, and rightly so. Yet, remarkably, they don't think of meat as food. Meat is luxury; food means carbohydrates. Nowadays they get their carbohydrates mainly by bartering with the farmers. In exchange for meat or honey or labor, they receive roots and grains: cassava, sweet potatoes, and rice. But cassava, sweet potatoes, and rice are all recent immigrants, domesticated food crops that have arrived in this part of the world during the last few hundred years. The Pygmy peoples who lived in Africa long ago must have had different sources of starch.

If you ask the Pygmies what they do when domestic starches are unavailable, you get a consistent answer. They go to the *kakbas* — these high, dry, microwoodlands — to find wild yams. A root. They have to get their root starches outside of the rainforest because the rainforest provides very little. In the relatively unseasonal world of the rainforest, most plants can grow all year round — and thus very few store energy in their roots. They don't need to.[8] But any plants living on the *kakbas* face the problem of water loss and seasonal water shortage, so they have evolved to handle the challenge by storing extra water and carbohydrates inside fat roots. That's why the foraging peoples of the Ituri, when they cannot get carbohydrates from another source, go to the *kakbas*. If the local farmers' crop fails, it's to the *kakbas* that the foragers turn. They know how to find the deep, large roots by looking for shoots trailing up from cracks in the rocks. They know exactly where the most productive *kakbas* are, and when to visit each to have the best chance of a good harvest. The Ituri Pygmies, in short, are a rainforest people using savanna woodland islands to provide their own fallback food.

Let's be clear about why it's interesting to think about people's foods in the African rainforest. It has nothing to do with special evolutionary relationships between human foragers and nonhuman apes. The Pygmy peoples of the Ituri are no more or less closely related to chimpanzees than the authors or readers of this book are. But their way of life is interesting because it shows how even when you live in a rainforest you can find woodland foods. And the one food they turn to is particularly interesting as well.

Unlike fruits, roots are not typical food for primates as a whole, since most primates live in root-poor forests. But for a woodland-living primate, eating roots is a logical adaptation. Indeed, when their favorite fruits and seeds are scarce, woodland primate species in Africa often survive by eating roots. Baboons, the most successful nonhuman primates of the savanna, eat roots extensively, though they concentrate on little grass corms and small bulbs rather than the large tubers that people often prefer. Digging for large deep roots apparently defeats even baboons. But woodland apes would presumably have dug at least as well as chimpanzees, and in hard ground they might well have used tools. Indeed, fragments of horns and long bones that have been found alongside fossils of woodland apes (the later "robust" forms from South Africa) have exactly the pattern of wear expected of digging tools. Bob Brain has found sixty of these putative tools, so many that he considers that "the digging of vegetable food from the ground was a long-lived tradition of particular importance in the economy of these early hominids."[9]

The teeth of fossil woodland apes, poorly adapted for eating leaves, look good for crushing roots. In fact, a classic blooper in the history of anthropology occurred when a broad and thickly enameled fossil tooth was identified as a woodland ape tooth, only later to be reassigned as a pig tooth. Pig teeth appear designed to chew roots.[10] The ancient habitats of woodland apes, moreover, must certainly have had high root densities, because fossil mole rats, animals clearly committed to root eating, are routinely found associated with fossil woodland apes. Mole rats today continue to live in all the African woodlands, but never in relatively rootless rainforest habitats.[11]

We think that woodland ape teeth were modified to eat roots, that roots were abundant in ancient woodland habitats, and that roots would have provided the key fallback foods when preferred fruits and seeds were out of season. So how did the shift in diet happen? Why, when ape teeth were inefficient at chewing roots and there were no traditions of finding buried treasures, did natural selection favor those innovative individuals who took the first steps toward root eating? How can we account for a single population of rainforest apes developing a root-eating habit that would be useful in the woodland?

We need a bridge, a step-by-step transition between the fruit-eating rainforest apes and the root-eating woodland apes. Modern chimpanzees have such diverse traditions and so many ways of solving problems that it would not be surprising if a chimpanzee community somewhere in Africa provided such a bridge. And in eastern Zaïre, within a chimpanzee group first studied as part of a tourist project sponsored by the Frankfurt Zoological Society, we find exactly the sort of behavior that could have set one little modest population of endangered rainforest apes on the road to the moon.

Tongo is a quiet forest. It is ordinarily visited by few tourists, despite being close to a main road and having wonderfully accessible wild chimpanzees who tolerate the presence of humans yet are never fed or touched by them. If that part of Africa were politically more stable, thousands of tourists would visit every year, bringing much-needed foreign currency into the country. Instead of tourists, however, in 1994 a million Hutu refugees from Rwanda moved into the region; and so the Tongo rainforest is now under heavy pressure from new settlement.

Few people, Africans or Europeans, have seen the chimpanzees that Annette Lanjouw habituated at Tongo in 1989, and no formal scientific research has yet been conducted there.[12] So it is to Annette Lanjouw that we owe the only description of the Tongo chimpanzee culture, which is special in one particular respect: It's a culture of water shortage. Tongo is unusually quiet because it has no rivers, no streams, no gurgling brooks or rushing falls. It is a rainforest island on an old lava flow, surrounded by dry bush growing in soils of younger lava. The central lava flow is perhaps five hundred years old, and now supports some fine rainforest trees, many of them figs. But step outside this core range, a few square kilometers altogether, and you enter a world hostile to chimpanzees: a bright, pale, scratchy, drought-adapted set of bushes and low trees. This surrounding hostile sea on the young lava must still wait many decades before the soils become rich enough to support a forest suitable for chimpanzees.

And even in the Tongo heartland, rain seeps quickly through the porous lava surface. There are no rivers, and not even any pools. So the chimpanzees have taken to an extreme the sponging tradition seen in some other wild chimp populations. Every day, says Lanjouw, the Tongo chimpanzees use sponges of moss to extract water lingering in tree holes, and the one day I spent there, I duly watched them do so. One might think they would do better to use their fingers, but sponging looks more efficient. I've measured chimpanzee leaf-sponges as holding about 10 cubic centimeters of water, whereas after dipping my hand in water I find that less than 4 cubic centimeters could be mopped off. And the time taken to lick is probably an important cost.

Vervet monkeys in dry parts of Kenya will spend up to 10 percent of their day "drinking" by dipping their hands, far longer than is good for them. I've watched the low-ranking individuals die from losing the competition for water.[13] Efficient ways to drink are very important if you have no standing water.

Water shortage appears to be the reason for a second Tongo tradition. Every now and again, a chimpanzee finds emerging from the blocks of volcanic cinder a peculiar stem or tendril that engenders excitement. Soon there's a frenzy of digging, with individuals plunging their arms up to the shoulders into the ground. And after many minutes the result of all this mad activity appears: Someone extracts a root. Competition for it breaks the social calm, producing squeaks and fights and appeals and comforts. As when chimpanzees kill a monkey, an adult male is first to possess this prize; and then others of both sexes and all ages press around and desperately beg for a piece. The root is divided and shared. Companions sit and chew or carry their pieces of the booty around with them, sometimes for hours as they move to new food sites.

Botanists were surprised when Lanjouw produced the particular root so valued by Tongo chimps, since no species had been known to have such a root. Identification is still uncertain, but it appears to be a *Clematis* that has become adapted to the special conditions of Tongo's low-water soils.[14] Its nutritional significance is still unclear as well. But Lanjouw describes it as very wet, and she thinks that for the wild chimpanzees it serves as a source of water. The root is a bottle. In a forest growing over porous lava, the tradition of digging and eating roots has appeared among wild chimpanzees as a local adaptation to water shortage.

Imagine, now, the world of the northeastern African forests 5 million years ago. Picture a population of apes in a rainforest like Tongo's, on a lava flow; or perhaps this forest grows around a *kakba*, as in the Ituri of Zaïre. The apes develop a root-finding tradition in the lava forest or the *kakba*, first as a way to get more water — in the style of the Tongo chimps. Then, as the result of a local drought perhaps, the little world of this group becomes completely cut off from the main ape populations. At the same time, its environment becomes progressively drier, so dry now that the usual fallback foods, the forest leaves and piths their ancestors relied on during drought, aren't available. If the apes continue to behave in the old manner, they will die. This isolated little group will become extinct. Luckily, however, they already know about roots, how to find them, and how to exploit them, based on their tradition of using roots as a supplementary water source. And the roots are there, all around them, because in this little rainforest on a lava flow or *kakba*, the plants have already adapted to seasonal stress. Given the luck of having the right kinds of roots available, the apes can increase their use of them, now not merely as a serendipitous source of water but as a critical cache of starch during hard times.

If the drought were to end relatively soon, these isolated apes might return to the richer forests, go back to their old reliance on fruits and seeds and meat, and reassociate with other members of the larger ape population living in contiguous forest. That may indeed have happened a number of times. But what if the drought persisted, so that instead of expanding, their little rainforest island continued to contract, becoming thinner and less productive until the threatened population of apes finally spread out in their search for food — for roots this time — onto the savanna woodland between the sparse rainforest patches? Here they would have found roots to be plentiful or at least sufficient, and they would therefore have discovered, for the first time in ape history, that they could survive beyond the wet forests.

Five million years ago our ancestors crossed the great ecological divide between tropical rainforest and woodland. While the old ape lineage continued relying on the forest in its traditional ways, conservative in behavior and morphology, natural selection favored rapid change for the new ape line on the woodland.

Already equipped with a way to survive, they took advantage of the open opportunity in front of them. The woodland apes kept their climbing abilities until some of them became early humans, and they continued harvesting fruits and seeds from trees when the natural orchards were productive. One line finally abandoned the old ape climbing adaptations around 2 million years ago. But they walked upright, so the bones and footprints show, by 4.5 million years ago.[15] Maybe it was worth the initial inconvenience of tiring bipedalism to carry roots to a tree where an individual could slowly consume them in safety. Tongo chimpanzees carry their roots for a kilometer or more, so perhaps root carrying was the habit that took the woodland apes upright. At the very least, we suggest, roots kept our ancestors from starving during times when the best foods, meat and fruits and seeds and mushrooms and honey, couldn't be found.[16]

With comparative rapidity, the woodland ape line branched several times, leading to species that probably covered the savannas and woodlands of Africa from west to south, and sometimes to two or more species sharing the same habitat.[17] Our own ancestors from this line began shaping stone tools and relying much more consistently on meat around 2 million years ago.[18] Their brains began expanding toward human size around 1.8 million years ago in an astonishing development that ended only half a million years ago.[19] They tamed fire perhaps 1.5 million years ago. They developed human language at some unknown later time, perhaps 150,000 years ago. They invented agriculture 10,000 years ago. They made gunpowder around 1,000 years ago, and motor vehicles a century ago. These are amazing events, alterations, and accomplishments. And yet, despite the extraordinary change that took place during our journey from rainforest ape to modern human, there has been continuity as well. We described earlier one continuity at the behavioral level: from the lethal intergroup raiding of modern chimpanzees, with their male-bonded territorial communities, to warfare among modern humans. And we raised the possibility that there is a biological foundation for these behaviors, in both chimpanzees and humans, that evolved before the ancestral split 5 million years ago.

Chimpanzee raiding and human warfare are not the same, however, so why should anyone imagine that they arise from the same source? How similar, or how different, are these two sets of species-specific behaviors? In becoming human (in turning into that upright-walking species who fights wars and makes peace, who gathers into communities and nations, who makes sexual bonds and breaks them, too, who uses language to expand intelligence and collect knowledge enough to design cathedrals and mousetraps and atomic bombs and myths), did we leave the old ape brain behind? Did we at some point simply jettison the whole thing as a worthless relic from the troubling shadow of time? Or is the elaborate, nervous and anxious and proud, superstitious and self-deceiving edifice of cerebral material that makes up our humanity still deeply infused with the essence of that ancient forest brain?

4

◄ ◄ ◄

RAIDING

THAT CHIMPANZEES AND HUMANS kill members of
neighboring groups of their own species is, we have seen,
a startling exception to the normal rule for animals. Add
our close genetic relationship to these apes and we face the
possibility that intergroup aggression in our two species has a
common origin. This idea of a common origin is made more
haunting by the clues that suggest modern chimpanzees are not
merely fellow time-travelers and evolutionary relatives, but surprisingly excellent models of our direct ancestors. It suggests
that chimpanzee-like violence preceded and paved the way for
human war, making modern humans the dazed survivors of a
continuous, 5-million-year habit of lethal aggression.

Until we face the evidence from chimpanzees, we naturally
imagine that warfare is a uniquely human activity. We might
see it as a practical way to control population density, or as an
outcome of specific cultural practices such as the invention of
weapons or an ideology of superiority.[1] We might stress that war
is based on calculation rather than instinct, and that it is an
instrument of policy.[2] Or we might see it more generally as the
product of social conditions. And even with the chimpanzee

evidence, the "blind instincts" of animal "hostility" seem far removed from the sophisticated calculations and ritual complexities that must surely lie at the heart of human war. From four hairy apes venturing across a valley in order to pummel some hapless neighbor to four hundred thousand flag-waving humans facing each other with guns and gas and rockets, tanks, artillery, electronic surveillance, and batlike bombers zooming overhead at two to three times the speed of sound: Can this bridge be crossed? Or is the similar pattern of violence among chimpanzees and humans merely a meaningless coincidence?

No human society offers a better opportunity for comparison in this regard than the Yanomamö, a cultural group of some 20,000 people living in southern Venezuela and northern Brazil in the lowland forests of the Amazon basin. Yanomamö provide the conceptual bridge here not because they are living fossils, which they aren't, but because they have been so remarkably protected from modern political influences. They are not aboriginal hunters and gatherers. For unknown centuries they have had agriculture, and for some time since the era of Columbus they have engaged in trade for metal axes. But they are nonetheless still the largest tribe on earth that has not yet been pacified, acculturated, destroyed, or integrated into the rest of the world. Though currently on the map of idealistic Western missionaries and cynical Brazilian gold miners, their traditional lands are protected thanks to Venezuela's creation in 1991 of a 95,000-square-kilometer biosphere reserve. The Yanomamö are not dominated by neighboring tribes, and their villages are scattered and mobile enough that no important intervillage hierarchies have developed. Each village exists alone, cut into the forest, in thrall to none other, permanently aligned with none, subject to no king or state or other external obligation. The Yanomamö village is a world unto itself, surrounded at a distance by many other such worlds.[3]

The Yanomamö are famous for their intense warfare, which has been described in great detail by the American anthropologist Napoleon Chagnon. Chagnon himself is a feisty and out-

spoken polemicist whose reports have generated some controversy among other anthropologists. Some think he exaggerates
Yanomamö war;[4] others are concerned that his reports, true or
false, are shocking enough that they might be used to justify
gold miners' incursions onto aboriginal land. But his data speak
as clearly as the people do. Yanomamö men call themselves
waiteri: fierce. Their villages are situated in the forest among
neighboring villages they do not, and cannot, fully trust, no matter how much they might wish to. Most of the Yanomamö people regard their perpetual intervillage warfare as dangerous and
ultimately reprehensible; and if there were a magic way to end
it perfectly and certainly, undoubtedly they would choose that
magic. But they know there is no such thing. They know that
their neighbors are, or can soon turn into, the bad guys: treacherous and committed enemies. In the absence of full trust,
Yanomamö villages deal with each other through trading, inter-
marriage, the formal creation of imperfect political treaties —
and by inspiring terror through an implacable readiness for revenge.

Their ordinary life is peaceful. Yanomamö are subsistence
farmers — swidden horticulturists — who acquire the bulk of
their calories from growing plantains and other fruits and vege-
tables in temporary plots. They are well fed. They need only
tend their cultivated plots for about three hours a day, and hunting brings sufficient meat. They expect and require little else.
They benefit from no encoded laws, written language, or count-
ing system beyond the number 2. Their clothing is, as Chagnon
phrases it, "decorative."[5] The men wear a string around the
waist that is tied to the foreskin of the penis; and it is a matter
of great embarrassment should the string come undone. The
women wear almost as little — a cord around the waist and an
apron of a few square inches — and are likewise decorous and
modest.

Villages comprise an average of ninety members, all related
to one another through male descent. That is to say, Yanomamö
men stay in the village of their birth while the women emigrate

before or upon marriage. Of course, as time passes and the population of a village grows, the blood ties dilute. Once village size reaches about three hundred, according to Chagnon's figures, the kinship center no longer holds. Some minor event or irritation leads to a squabble, then a fight, and finally a complete rift. The entire village divides, roughly along male kinship lines. The two new villages, perhaps separated only by a small river or a piece of swamp or stretch of forest, remain friendly at first. But over time the friendship recedes. Tension builds. Eventually war becomes likely or happens, and so the two villages move farther away from each other, deeper into the forest. The process recalls the separation of the Kasekela and Kahama chimpanzee communities.

Intervillage war, declare the Yanomamö, takes place not over resources. It may be unleashed by something as theoretical as suspicion of sorcery or as mundane as a trivial argument. Or a couple of men from different villages may begin fighting over a failed agreement, sexual jealousy, or suspicion of adultery. Yanomamö say it happens most often over women. In any case, tradition allows for such small conflicts to be settled by a few formal fighting games, such as a chest-pounding duel. One man offers his chest to be struck by the other, takes as many blows as the other cares to make, and then returns the same number. If both parties are satisfied, the conflict can end there. If not, a side-slapping duel ensues, with blows dealt to the ribs by an open hand or the flat side of an ax, risking serious injury. Side-slapping duels can end the conflict, or they can escalate further into club fights, where the men clobber each other over the head with eight- to ten-foot-long building poles. A club fight is the last level of formality. If there is no peace after that, the village men assemble themselves and move on to war.

Yanomamö fight their wars with two combat styles, one uniquely human. *Nomohori* — the dastardly trick — speaks for itself. Here's one way it can happen: The men of a village invite their supposed allies to a feast. They treat them so well that the guests relax utterly, lying down in the hosts' hammocks. And then, in unison, the hosts turn and attack: split skulls open with axes, beat their guests with clubs, pelt them with arrows. Men are slaughtered, women and girls taken captive. One wonders why anyone would ever go to a party at another village.

The second technique of war is the *wayu huu*, or raid. Yanomamö war raids begin in a conversation, with a party of ten to twenty men agreeing to kill some particular enemy. They make an effigy of that person out of straw or painted wood and fire arrows into it — though in fact they will be satisfied to kill any man from the enemy village. After an evening of ceremonial and emotional preparation, the raiding party leaves the next morning. Villages actively at war with each other tend to move far apart so that a raid may require walking four or five days through the forest before reaching the enemy village. During this walk, some of the men may quit the party, complaining of illness or injury; but men who never follow through on raids risk becoming known as cowards and having their wives considered fair game for seduction. Upon reaching the outskirts of the enemy village, the raiding party waits quietly overnight. At first light they split up into two smaller parties, four to eight men each, and wait quietly in ambush for a lone victim, hoping to surprise someone sleepily off guard, out to urinate or fetch water perhaps. If they should find, instead of one man, two or more men capable of defending themselves, the raiders will fire a volley of arrows into the village and run. But if they find a lone enemy, they shoot him with deadly curare-tipped arrows. The raiders then immediately flee, anticipating a chase and ultimately a retaliatory raid against their own village.

The stated object of a raid is to kill one or possibly two men and escape. If the raiders can do so without risking losses, however, they may abduct a woman from the enemy village. The abducted woman will be raped by all the raiders, taken to their village, raped by the remaining men in the village, and then given as a wife to one man. She can expect to spend the rest of her life with her new companions.

How common are these startling practices? Yanomamö men

who have killed or shared a kill must undergo a purification ritual called *unokaimou*. They are then known as *unokais*. About 40 percent of Yanomamö adult men are so honored, Chagnon reports, and some of the most prominent men in Yanomamö society have killed often, including one well-respected man who has been purified for killing twenty-one times. Although some critics have pointed out that in a typical killing two or three men will fire arrows into the victim, thus creating more killers than victims and so exaggerating the importance of *unokai* statistics,[6] Chagnon has also compiled data on victims. Some 30 percent of all Yanomamö men die from violence.

Raiding may seem a futile activity, but like military heroes around the world, Yanomamö *unokais* are honored by their societies and ultimately rewarded. Because Yanomamö culture allows polygyny, the rewards can directly translate into reproductive terms. *Unokais*, so Chagnon discovered through analyzing data from several villages, have more than two and a half times the average number of wives as other men, and more than three times the average number of children.[7] Lethal raiding among the Yanomamö, it seems, gives the raiders genetic success.

As subsistence farmers, of course, Yanomamö are not typical of humans at the trailing edge of the Pleistocene, before agriculture was invented. But then human society varies too much for any single group to be perfectly representative of humans at any stage or state. The people studied by Chagnon aren't even representative of all Yanomamö — as his critics have been quick to point out. Still, Chagnon has traveled widely through their land, studying the cultural styles of several villages located in distinctively different ecological situations. His data, drawn from several communities during five years of study altogether, provide enough reference points to allow for an extended comparison between the chimpanzee system of lethal raiding and a human system of war in a society where intervillage political autonomy resembles the intercommunity isolation of chimpanzees.

Yanomamö people and Gombe chimpanzees. How are primitive war and lethal raiding different among humans and chimpanzees, and how are they similar? The differences are easy to describe. Even here, in one of its simpler forms, human warfare is hugely different from chimpanzee raiding and in every way more complex. War among the Yanomamö is an overtly acknowledged relationship, part of an escalating tension between villages, possessing a history that men and women discuss. It can be provoked by sorcery. It can be motivated by revenge. The combatants prepare ceremonially. They use hand-held weapons instead of teeth, and their poisoned arrows can pierce the body of an individual or be fired in a volley against a whole village. Their war can include dastardly tricks. It sometimes has a plan. It targets specific enemies. A raid often takes days, not hours. Abduction and rape are common. Retaliation is expected. And so on. When Gombe and Yanomamö are compared, the gulf that divides our two species is unmistakable. Because language makes discussion and meaning possible, the cultural dimensions to human war will always make it richer, more complicated, more exciting, as well as more self-deceiving and confused, than chimpanzee intercommunity violence.

But a human inventiveness that elaborates the deceptions and meanings and possibilities of warfare needn't blind us to the common threads. Like Yanomamö villages, chimpanzee communities are kinship groups based on aggregates of closely related males and unrelated females who have emigrated from other kinship groups. Yanomamö villages vary in size from forty or fifty to around three hundred; chimpanzee communities range in size from twenty to around one hundred ten individuals. Like Yanomamö war, chimpanzee lethal raiding takes place when a subgroup of males — in both cases, gangs of roughly a half dozen to a dozen — deliberately invades the recognized territory of a neighboring community.[8] And the style of the raid is similar as well. Remember what happens among chimpanzees.

The ape raiders are quiet, alert to enemies. If they find a neighboring party that includes enough individuals to constitute a threat — two or more adult males, for example — they will retreat, sometimes after making noisy displays. But if they find a lone neighboring male, an infertile female, or a male-female pair, the group will stalk and then attack brutally, sometimes lethally. The raiding chimps appear to assess tactical risk by locating and observing their enemy before attacking, they make sure they have a clear numerical advantage, and they try as well to gain the advantage of surprise. In addition, such attacks typically immobilize the victim, so that the attackers themselves are barely injured. The victims may be either male or female; but the aggression usually focuses more severely on adult males, less severely on obviously fertile females. Young females at the start of their breeding careers (nulliparous females) are most likely to escape injury and can be forced to travel back with the raiding party into their home territory.

Based on the evidence of the chimpanzees' alert, enthusiastic behavior, these raids are exciting events for them. And the mayhem visited on their victims looks a world apart from the occasional violence that erupts during a squabble between members of the same community. During these raids on other communities, the attackers act as they do while hunting monkeys, except that the target "prey" is a member of their own species. And their assaults, as we have seen, are marked by a gratuitous cruelty — tearing off pieces of skin, for example, twisting limbs until they break, or drinking a victim's blood — reminiscent of acts that among humans are regarded as unspeakable crimes during peacetime and atrocities during war. In Gombe about 30 percent of adult male chimpanzees died from aggression[9] — about the same percentage Chagnon found in the Yanomamö villages he studied.

The differences are important, but so are the similarities, because they hint at a shared cause and common origin. The Yanomamö suggest to us that as human economic and ecological conditions move closer to those of chimpanzees, so the patterns of violence in our two species start to converge. The conditions that make Yanomamö society similar to that of chimpanzees are their political independence and the fact that they have few material goods, no gold or valuable objects or foods to fight over. In this relatively stark world, some of the more familiar patterns of human warfare disappear. There are no pitched battles, no military alliances, no strategies focused on a prize, no dramatic seizure of stored goods. What remains is the dastardly trick and lethal raiding, penetrating expeditions in search of a chance to attack, to kill a neighbor, and then to escape.

Among the Yanomamö we see the simplest, most rudimentary form of human warfare that has been thoroughly described. Unfortunately for anthropology, much less is known about warfare among equivalently isolated foraging people, people who could give us insights into life without agriculture. The problem is that no contemporary hunter-gatherers are as numerous or as free as they used to be. They all endure within larger cultural and legal systems. The best accounts therefore come from early travelers, but these tend to be quirky and often disreputable sources. How far can we trust William Buckley, for instance? Buckley was an English convict in Australia who escaped to live among Aboriginal foragers from 1803 to 1835. He described numerous violent deaths, including a raid in which a man was speared in the hut where he himself was sleeping.[10] Buckley's extended accounts of battles and raids and fights by men over women ring true, and we are tempted to take his word if only because Australia was a whole continent without agriculture until the Europeans arrived. But the distant reminiscences of an escaped convict and adventurer have to be regarded with some suspicion.

So we come back to the Yanomamö. Do they suggest to us that chimpanzee violence is linked to human war? Clearly they do. The appetite for engagement, the excited assembly of a war party, the stealthy raid, the discovery of an enemy and the quick estimation of odds, the gang-kill, and the escape are the common elements that make intercommunity violence possible for both. Language among the Yanomamö makes other things possible, like planning, formality, ritual, and treaties. But it is not needed for the violence itself, the carrying out of the lethal raid.

Lethal raiding is not only one of the principal styles of Yanomamö war, it has also been one of the commonest styles of primitive war throughout the world.[11] Harry Turney-High, an American sociologist and anthropologist who served as a military officer on horseback during World War II, remarked in *Primitive War* (1949) that the social scientists of his day had largely failed to examine or comprehend the single most important set of behaviors occurring in virtually all human communities: war. Primitive people, according to Turney-High, killed each other as eagerly and ordinarily as anyone else, but they did not fight what he chose to call "true" or "civilized" war. "Despite their face-painting and sporadic butchery," these people nonetheless existed "below the military horizon" and perpetrated something characterized by disorganization and a lack of planning, discipline, and tactics. More like gang murder than war. They were culturally incapable of participating "in a fight which could be called a battle."[12]

Turney-High borrowed from his American cavalry manual a list of fourteen tactical principles that for him defined the art of true war. Number 8 on that list is the principle of surprise, which also happens to be the single principle he found most characteristic of primitive war. Native American tribal groups as a whole, for example, used surprise as their central tactic. Nineteenth-century Dutch commentator David DeVries wrote of the Leni Lanape of Delaware, "As soldiers they are far from being honorable, but perfidious, and accomplish all their designs by treachery; they also use many stratagems to deceive their enemies and execute by night almost all their plans that are in any way hazardous." Though moralists like DeVries described the Native American approach to war as "treachery," Turney-High noted that the Europeans themselves were often equally treacherous. "Most of the Dutch victories over the Indians were accomplished through broken promises and the mass murder of undefended villagers."[13]

Studies and observations since the publication of *Primitive War* have confirmed, in example after example, the importance of surprise as the central tactic among primitive people. Many are strictly foraging societies, with exotic names like Aleut and Andamanese, Shivwits and Squamish.[14] Surprise — the ambush, the trap, the sneak attack, the midnight raid — has been a fundamental fighting tactic of tribal groups from North America to Africa, Europe to Melanesia.

When you build on these simple elements with distinctly human complications, more complex styles of war start to emerge. In highland New Guinea, for example, remote tribes like the Mae Enga have little in the way of material goods, and they use weapons as simple as those of Yanomamö. But they live at a high density, up to more than one hundred people per square kilometer,[15] while the Yanomamö are spread out at a chimpanzee-like density of less than one person per square kilometer. As a result of their comparative crowding, the Mae Enga clan parishes (dispersed equivalents of villages) are able to form stable alliances with one another. When allied Mae Enga groups meet for a war with other allied groups, they sometimes assemble and participate in ritual battles that look like an exaggerated version of a sports day. The battles are normally peaceful, little more than two lines of men running at each other from a few yards apart, with the action called to a halt if someone should be wounded. This courteous assemblage of opposing sides, lined up and mentally measuring each other, makes Mae Enga warfare look very different from chimpanzee raiding. But once every decade or so, one alliance turns out to be enormously stronger than the other, a result of diseases or defections on the weaker side. With power no longer balanced, the laughter stops. The larger alliance destroys the opponents, leaving many dead. The survivors must leave their land forever, and live as refugees with other kin.

So a society like the Mae Enga elaborates on the Yanomamö theme, with a style of warfare more distant from the simple pattern of chimpanzee violence. Mae Enga use ritual war as one peculiar, particular way of estimating numerical odds — what a chimpanzee gang or a Yanomamö war party does when it finds either one or several of the enemy and then quickly calculates whether to attack or retreat. Mae Enga warfare is different: more obscure, more complicated. But the continuity can still be traced. Bring on the flags and uniforms, the horses, armor, guns, ships, and tanks — alliances and weaponry multiply geometrically the complications, but some fundamental bridges to our past remain.

If all primitive peoples behaved quite like the Yanomamö, our case would be strengthened. But styles, frequencies, and intensities of warfare differ widely even among primitive peoples, some of whom lead decidedly more peaceful lives than the Yanomamö. The picture remains ambiguous, and it leads some anthropologists to argue that the chimpanzee pattern, interesting though it may be, is not directly ancestral to the human one.[16]

What intrigues these anthropologists is partly that the best modern representatives of our preagricultural past — foraging or hunter-gatherer societies — don't seem to show the expected level of violence or warfare. In general, hunter-gatherer people evince some of the most delightful and admirable ethics found anywhere. They may possess only a few rough and worn objects and little food beyond what is about to be eaten, but whatever one individual has is usually shared. People cooperate, and they promote cooperation. When one man tries to make himself better than his fellows, he is scorned, so that no one can become the "big man" or a petty tyrant over others. Hunter-gatherer societies are capable, anthropologists agree, of an "extreme political and sexual egalitarianism."[17] But it is important to remember that egalitarian is not the same as peaceful. And even if it were, the sharing and communal ethos of hunter-gatherer societies only involves relationships within a community group, and in trying to trace the evolution of warfare the important question is not how hunter-gatherers treat others within the same group, but how they treat those from other groups.

The answer is unclear. All modern foraging cultures have been affected in important ways by recent cultural changes, whether by trading food with agricultural people or being forced to live in marginal areas or being dominated, parceled, and pacified by a powerful nation-state. In addition, all the historical records of foraging cultures are poor. But we do know that no truly peaceful foraging people has ever been found or described in detail.

Statistics challenge the notion of the gentle forager. A global assessment of the ethnographies for thirty-one hunter-gatherer societies found that 64 percent of them engaged in warfare once every two years, 26 percent fought wars less often, and only 10 percent were considered to fight wars rarely or never.[18] So the record suggests regular, almost constant war for most foraging cultures. As for the 10 percent that appear to be exceptions, one wonders how long "never" really is, with the ethnographic record so sketchy and our time frame so narrow. An anthropologist settled into Western Europe for the two decades between 1920 and 1940 might report a Germany, France, Poland, and Czechoslovakia at peace. A couple of decades or a few generations of peace tell us little about history and far less about evolution.

Yes, peaceful foragers have been repeatedly hoped for. Thus many anthropologists were excited by the discovery in 1971 of a hunter-gatherer enclave in one remote part of the Philippine Islands. A man named Manuel Elizalde described a group of twenty-six people who said they were the Tasaday and who looked to be the very prototype of peaceful primitives. Elizalde went on to claim that the Tasaday wore clothes made out of leaves and had their own unique language — with, most remarkably, no words for conflict, violence, weapons, or war. The Tasaday became famous in the Western world, featured, for example, in *National Geographic* articles and a 1975 bestseller, *The Gentle Tasaday*, written by journalist John Nance.[19] For a decade of European and American college students the Tasaday came to symbolize all the wonder and delight and peace of being primitive. But then, in 1985, a Swiss journalist, Oswald Iten, pronounced them a hoax. Instead of being aboriginal foragers, the Tasaday, Iten claimed, were poor local farmers paid to participate in a gigantic fraud run by Elizalde in league with President Ferdinand Marcos. The two men used the existence of these purported hunter-gatherers as an excuse for setting up a preserve that enabled them to maintain exclusive rights (for mining, timber, and so on) over the Tasaday's large "homeland." Yet the original picture of the Tasaday had been so appealing and, for many, so convincing. Was Elizalde right? Or was Iten right? A terrific academic wrangle ensued. Eventually, all independent investigators agreed that the Tasaday were not an isolated people at all. Instead, they were local traders, depending on agricultural foods and tools made from cultivated bamboo. Their stone tools were fraudulent, their leaf clothes a show, and their language differed from that of their island neighbors at most by accent. Their Pleistocene innocence was an invention of unscrupulous exploiters, schemers from the urban elite of a modern nation who knew exactly what would sell to the outside world.[20]

Peaceful primitives have been conjured in more genuine cultures. *The Harmless People*, Elizabeth Marshall Thomas's classic description of life among the !Kung San hunter-gatherers of the Kalahari, shows how wonderful they often are, quite as the Tasaday were supposed to have been. But raiding has been described among the !Kung San.[21] In revenge for a killing, a party from the victim's group located the killer's group and attacked them early in the morning as they slept, using arrows and clubs to kill everyone they could: men, women, and children. And violent death rates among the !Kung, we now know, exceed the murder rate recorded in contemporary America's worst cities.[22] Life among the harmless people is rawer than it first appears.

It's not just among Yanomamö that primitive war can be deadly. Anthropologists have occasionally been able to gather statistics on warfare among independent peoples uninhibited by intervention from more powerful tribes or governments. Violence accounted for the deaths of about 19.5 percent of adult men among the Huli of highland New Guinea; for the Mae Enga and the Dugum Dani, also of highland New Guinea, warfare produced adult male mortality rates of 25 percent and 28.5 percent, respectively. For the Murngin of Australia, the figure was 28 percent.[23]

Even though there must be primitive societies with far lower rates of killing, the figures suggest that modern life is on the average less violent. There are obvious reasons why this might be true. A tribal community like the Yanomamö, lacking a judicial system, has no formal mechanism for punishing crime, so that any punishment is imposed by the victim's relatives, which, of course, often leads into a cycle of blood feuds. Moreover, since in the primitive world the edge of your allegiance extends to the outskirts of your village rather than to the borders of your nation, war may be more quickly induced by minor local events and then more easily perpetuated.

We know well that social patterns can change with great speed. Consider the Yanomamö. Like tribal people everywhere, they are losing their old way of life. Increasingly, and in spite of Venezuela's recent formal protection of a large piece of their homeland, global economic interests, missionaries, and in some places gold miners and government agencies still are pressing into the Amazon and pacifying the Yanomamö. While forsaking their cultural certainties and self-assurance, adjusting to new legal controls, and being devastated by strange new diseases, many Yanomamö are simultaneously becoming less violent. Introduced laws and ethical codes, able to end the seemingly endless cycles of blood revenge that until recently characterized relations between villages, testify to the good things that can come from acculturation to the modern world.[24] It may be quite soon that anthropologists can describe them as "never" having wars.

Even more striking, partly because the process seems even faster and closer to completion, is the case of an aboriginal people living several hundred miles to the southwest of the Yanomamö. At the base of the Andes, within a vast area of rich and deep-valleyed rainforest, the Waorani hunt small game with blowguns and poison darts, large game with palmwood spears, and cultivate sweet manioc and plantains as they have done for as long as anyone can remember. These people, only about five hundred individuals altogether, were until the late 1950s settled thinly across their land, roughly divided into four mutually hostile communities whose members spoke slightly different dialects of the language Auca. Each of the four communities was further subdivided into smaller settlements that were continuously raiding and killing one another and simultaneously raiding and killing the indigenous peoples who lived on the other side of their territorial boundary. The raids conducted internally between Waorani settlements were generally retaliations — over disagreements in marital arrangements or for suspected sorcery — blood feuds perpetuated by revenge for previous killings. A typical raiding party would break into a neighbor's house during the night, surprise the sleepers, and slice up as many people of both sexes and all ages as possible. Then the raiders would disappear into the forest, abandon their own settlement, and move many miles away in order to hide themselves from the anticipated return raid. When Waorani crossed their territorial boundary and raided households of their Quichua neighbors, however, their motivation may have been less revenge and terror than theft; they would take valuable metal tools and sometimes kidnap Quichua women. The Quichua, in turn, sometimes crossed the boundary to raid Waorani households and kidnap Waorani women and children, who until the

1950s might be sold into forced labor on haciendas in the Andean foothills.

Though their Quichua neighbors often had access to shotguns, the Waorani acquired such a terrifying reputation that with a population of only five hundred, armed solely with wooden spears, dispersed and fearing each other, they had nonetheless carved out a 20,000-square-kilometer traditional homeland in the rainforest. That reputation was earned, however, by a life circumscribed by suspicion and feud, by the constant fear of violent death and the actuality thereof. The rate of violent death among the Waorani was calculated to be an astonishing 60 percent.

In 1958 the American Protestant missionaries came, for several years making their presence known with the buzz of a single-engine plane overhead, dropping down to the surprised and curious people important gifts of tools, clothes, and food. Missionaries next established contact on the ground. Then, using loudspeakers to promote communication between feuding groups, they ended the blood vendettas and encouraged many of the Waorani to settle into a central village, with infirmary, school, church, and landing strip. Around the same time, the Ecuadorian government proclaimed a tenth of the Waorani traditional land to be protected as a reservation. Most of the widely scattered Waorani moved onto this land, and the missionaries encouraged them, through persuasion and example, to desist from such traditional sins as killing one's neighbor and having more than one wife.

Anthropologists Clayton and Carole Robarchek describe the Waorani as having been engulfed in a self-perpetuating cycle of offensive-defensive raiding of uncertain historical origin — a "more or less stable balance of terror with constant raiding among the various social groups" — broken only by the intervention of an outside cultural force in the person of the American missionaries.[25] The missionaries, about a half dozen women and one man, persuaded them to end their internal and external raiding by providing the "social mechanisms" that allowed friendly contacts between enemies and encouraging the development of trust. Waorani peace also brought such benefits as medical care, schooling, metal tools, shotguns, flashlights, and the like.

Here is a remarkable change: from arguably the world's most bellicose tribe to a rather pacific group of people clustered on a government-demarcated reservation one-tenth the size of their previous homeland, a people who are still apparently content, even perhaps eager, to live in a new way and acquire the material benefits of the Western economy. Though twenty years is a short time, much too soon to know what the change means, perhaps it will come to stand as proof, as the Robarcheks conclude, that the age-old habit of killing each other in lethal raids ended among the Waorani "because the people themselves made a conscious decision to end it."[26] For anyone who imagines people to be biotic robots incapable of making significant changes in their lives, the story of the Waorani could serve as a challenging counterexample.

It does not, however, demonstrate that the Waorani, as they are led away from their aboriginal lives into something resembling contemporary Western culture, are entering a world devoid of violence. The American Protestant missionaries who initiated this change among the Waorani, after all, come from a society notorious (among industrialized nations) for its high murder rate; and the larger Western culture they represent, in spite of all its ethical standards and good intentions, has during the last three generations strung a bridge of barbed wire and blood from the Somme to Sarajevo and hosted the two largest, most deadly, and destructive wars in global history. If the Waorani someday do become fully Westernized, they will have traded a life marked by the flight of a palmwood spear for one measured by the parabola of a ballistic missile.

It is clear that a few societies have indeed managed to avoid outright war for extended periods. Switzerland may stand as the best modern European example. The Swiss have maintained their military neutrality since the seventeenth century, were officially recognized by other European powers to be neutral in 1815, and, in spite of a near entanglement in the Franco-Prussian war of 1870–1871, have avoided wars ever since — this on a continent wracked by wars. But the Swiss retain their peace by maintaining a large army, compulsory military service for all young men, live mines buried at critical bridges and passes, and deep protective bunkers carved into the mountains and stocked with enough food, water, and other supplies for the full army to withstand an extended siege. The Swiss are also, of course, effectively isolated from their neighbors by the Alps.

A survey of the ethnographic record for fifty representative non-state groups from around the globe found four of those societies — the Toda, Tikopia, Dorobo, and the Copper Eskimo — to have had no regular military organizations or warrior class, apparently as a consequence of finding themselves situated, like Switzerland, in extreme geographical isolation from their neighbors. Nonetheless, individuals from all four groups did fight and kill outsiders when the occasion demanded.[27]

Then there is the case of the Semai Senoi, an aboriginal people of about 13,000 who live in communities of up to 100 individuals, cultivating gardens and hunting meat in the rainforests of peninsular Malaysia. They must be considered, according to one anthropological report, "among the most peaceful people known."[28] No one appears to know much about the history of the Semai, however; and so some speculation is required about their lives before Malaysian national law, politics, and economics pressed a significant imprint onto their fragile world. Perhaps they were by choice a peaceful people; but they look more likely to have been merely disempowered, and history doesn't tell us for how long they lived without violence.

Our examination of the anthropological record so far is sobering. How extremely common war appears. How extraordinarily rare (once we look beyond the narrowest time frame of, say, one to three generations) any lasting state of peace seems to be. And the comparison between human war and chimpanzee raiding remains unsettling.

But what do the parallels prove? Chimpanzees, however bright they might be, however interesting and clever and entertaining, are animals. Animals don't have art, music, literature, and traditions and systems of ethics and religion and ideals. In short, animals lack human culture. Isn't culture the thing that gives us wings, that frees us from a slavery to passion and the violence of a tooth-and-claw nature? Surely culture makes us who we are. And isn't it obvious that humans show an astonishing variability in their social systems, proof positive that humans can simply "invent," through culture, virtually any reality and style of being and doing they wish?

This sort of question is so important that it compels us to digress from the specific comparison with chimpanzees. We have already touched on the issue. Now we need to consider more directly the question of what makes people tick. One view is that people can freely invent their societies and styles by cultural choice. If that's right, then presumably we ought to find samples of human societies that demonstrate the entire gamut of possibility. Yes, of course, there are repressive, aggressive, bellicose societies. We have already examined several; perhaps we live in one. But there should be peaceful ones, too.

Where are those exceptions to the general rule of human bellicosity, those wonderful places where people are not only at peace with each other and with their neighbors, but also at peace inside, in their own hearts and minds? They are hard to find in the world today. Is it possible that Western civilization has already etched its corrupting influence so far around the globe that our perceptions are now entirely distorted? Perhaps until recently there were many primitive Gardens of Eden, attractive little places where Western culture had not yet dropped its contaminating fruit. In short, have there been genuine human paradises existing until recently, special places where special people absorbed in special cultures constructed their own excellent worlds and simply chose peace and happiness?

5

, , ,

PARADISE IMAGINED

THE SEARCH FOR PARADISE is at once the quest for a real place and a journey to distant islands of the mind, places representing the potential for human perfection. Atlantis. Eden. Elysium. The bower of the Golden Apples. Paradise is a favorite theme of cultures around the world, and the idea has often operated by idealizing the rough reality of an actual landscape. By the time of ancient Rome, for example, the poet Virgil was writing of a mythical paradise for young shepherds in Arcadia, a backwater district of Peloponnesian Greece that, a thousand years earlier, had survived untouched when the rest of the peninsula was overrun by Dorian invaders.

Postclassical Europe developed its own versions of paradise, and by the late Renaissance Europeans were seriously debating whether the newly discovered American continents represented a real-world expression of the ancient fantasy.[1] But by the nineteenth century, much of the American landscape was already tainted by a mundane familiarity, so many people from both sides of the North Atlantic turned their hopes toward the South Pacific, a warmish place still sufficiently remote and unexamined to harbor any number of appealing images. To this day, visions of a paradise in the South Seas remain alluring motifs in Western popular culture, appearing not just in cheap advertisements promoting escape and romance through island vacations or in B-grade movies but in a serious way in major works of art and literature and even anthropology.

These contemporary images of paradise, no matter how truly or falsely they may represent their subjects, remain important because they project a particular, widely accepted view about human nature. Many of us who have seen the paintings of artists like Paul Gauguin, read authors like Herman Melville, and absorbed the ideas of anthropologists like Margaret Mead, find deeply comforting their evocation of paradise and their notion that human evil is a culturally acquired thing, an arbitrary garment that can be cast off like our winter clothes. It's a seductive vision, and in the hands of these talented figures it has been expressed with drama and conviction. It appeals because it gives the impression that if only we could get things right, we would locate the perfect world. It stimulates us to do good works. It offers hope amid the gloom.

Optimism is a wonderful emotion. But the vision of paradise that comes from the balmy islands of the South Seas is flatly challenged by the ubiquity of warfare and violence across time and space. How could these figures of genius, Gauguin and Melville and Mead, have gotten it so wrong? Or perhaps they were right after all. As we disentangle the arguments that brought them to their own individual but shared images of paradise, we find a remarkable thread unwinding that connects all three. Each found paradise in the same special way. Each imagined paradise as a place without men.

For artist Paul Gauguin, the South Pacific was a place of strong light and high contrast, with rich primary colors and readily available nude models. Gauguin was both a practitioner of French Impressionism and an inheritor of French Romanticism. And in his brilliant oils of Tahiti the artist painted Rousseau's noble savage with light poured and puddled like liquid. The

noble savage was specifically female and vaguely Christian, an
Eve of the tropics, produced and reproduced in an entire series
of pensive, self-possessed, and yet curiously provocative nudes
showing serenity, ease, sexuality, and freedom. In the painting he considered his masterpiece, entitled
Where do we come from? What are we? Where are we going?
Gauguin summarized most fully and dramatically his romantic
vision of paradise. Painted in December of 1897, this massive
canvas — four and a half feet high and well over twelve feet wide
— presents a pleasing scene of women and girls in repose, clustered in three groups before a stream in a grove of trees, crossed
by long shadows and spottily illuminated by a warm and orange
light cast from the setting sun. Intermingled with the human
figures are contented domestic animals. One girl, preadolescent,
mildly androgynous, stands at the center, sun-brightened, her
arms raised, poised to pluck a ripe fruit from a tree. But there are
shadows on either side. To the right a gloomy pair plans or plots
— as if the expulsion from Eden has already begun. To the left a
white-haired crone sits, face in hands, her eyes closed. She could
be fading toward death. Here is a moment in primitive time
when humans and animals lived harmoniously within nature's
garden, not yet overcome by sorrow and pain, time and death. It
implies that the real-life present moment in Tahiti might somehow represent an existence near, or near enough, to paradise.

"I do believe that not only is this painting worth more than
all previous ones but also that I will never do a better one or
another like it," Paul Gauguin wrote in a letter to his agent in
Paris.[2] But his contemporaries disagreed. The canvas was nailed
inside a wooden crate and shipped to Paris, where it fetched, in a
package deal with seven other paintings, 1,000 francs. A disap-
pointing sum. Nowadays, of course, it's worth a bundle. It hangs
in Boston's Museum of Fine Arts, testimony to the beauty and
power of the romantic vision.[3]

Back in Tahiti, however, real life was far from romantic.
While he was painting *Where do we come from?* Gauguin struggled in a deep depression, overwhelmed by the recent death of

his daughter Aline. As soon as he finished his masterpiece, the artist walked alone into the mountains, took an overdose of arsenic, and lay down to wait for death. But he took too much. So he vomited it up, spent three miserable days staring at the painting from his bed, and then recovered to continue a lonely life of battling an unappreciative public, hostile officials, and unsympathetic compatriots among the island colonials.

Still, there was a paradise — of sorts. When the artist first came to Tahiti, so he claimed in his book, *Noa Noa*, he found sexual opportunity everywhere. The only problem was that, according to the Tahitian men, the adolescent girls and young women of Tahiti wanted to be raped. "I saw many young women with untroubled eyes; I guessed that they wanted to be taken wordlessly, brutally. A desire for rape, as it were. The old men said to me, speaking about one of them: '*Mau tera.*' ('Take this one.') But I was timid and could not bring myself to do it."[4] He didn't need to. As an exotic and well-connected outsider, the artist himself was a promising catch. He quickly took on an adolescent girl companion, the lovely Titi, whose attentions he treasured for a while. She was a city girl, however, half white and corrupted by too much contact with Europeans, which meant she would "not serve the purpose I had in mind," whatever that was. In the country, where Gauguin soon went, he hoped to find "dozens of them," but, so he worried, "they would have to be taken in Maori style (mau = to seize)." His concerns were unnecessary. Before too long a thirteen-year-old girl was given to him by her mother to be his wife. And though he lived in great economic poverty, Gauguin's sex life continued to be rich and arrogantly expressed. He outraged the colonials by decorating his home with startling pornography; and he took a succession of young lovers, though he eventually found it hard to overcome the reluctance of girls put off by his venereal lesions.

Toward the end of 1901, Gauguin was so besieged by the local authorities that he decided to seek, as he wrote, "a simpler country with fewer officials."[5] He took a boat to the Marquesas Islands, where he bought a small plot of stony land from a Catholic mission and built on it a modest hut. He died there, impoverished, unhappy, and about to serve a three-month prison sentence for slander, on May 8, 1903.

For Paul Gauguin, then, the reality of life in the South Pacific was difficult and distressing. But the paradise he coated his canvases with remained serene, sultry, and exclusive. For him, the ideal of the South Seas was his own private club undisturbed by the presence of other men, packed with girls and young women who were simultaneously innocent and available. It was an island with only one man in residence: creator and voyeur at once, gazing at an oiled dream of nubile young women and pleasing if naive concepts of a peace in nature.

A generation before Gauguin painted his Tahitian Eves at the edge of paradise, a young American writer named Herman Melville created his own compelling vision of paradise in the South Seas. Whereas Gauguin was to clutch at fame in Tahiti and finish his life bound for jail in the Marquesas, Melville garnered fame through an adventure in the Marquesas and wound up in jail in Tahiti. These days, Herman Melville is famous primarily for his encyclopedic whaling epic, *Moby Dick*. But his book about the Marquesas, *Typee*, was not only Melville's first book, it was his biggest success during his own lifetime and for a half century after, influential not merely as a gripping tale but also as a work of ethnography.[6]

The writer entered the South Pacific as a common seaman aboard a 358-ton whaling vessel, *Acushnet*. The *Acushnet* rounded the horn of South America in April 1841 and by June had traced the western coast as far north as Peru. In pursuit of whales, the ship turned away from the continental edge at Peru, passed through a volcanic archipelago called the Galápagos Islands, and then caught the trade winds out into the deep South Pacific.

Melville's adventure in the Marquesas began soon after the *Acushnet* anchored in Taiohae Bay of Nukuheva, largest island of that group, where he jumped ship. The Marquesas are an archipelago of ten volcanically formed islands roughly 4,000 miles west of the Galápagos and around 850 miles northeast of Gauguin's Tahiti. They were colonized a century or two before the birth of Christ by Polynesian people sailing east from Samoa across 2,000 miles of open sea to establish patriarchal yet relatively egalitarian societies in which every first-born male was called *haka-iki* — chief — and might potentially acquire high status through war, wealth, or politics. Marquesans lived within small communities separated from each other by geography and a perpetual state of suspicion, hostility, and warfare. Warfare ordinarily amounted to a regular series of clashes, ambushes, and raids for the purpose of acquiring bodies to be eaten ceremonially and also in retaliation for an enemy's ceremonial cannibalism.[7]

That cannibalism, combined with an open and comparatively uninhibited sexuality, made the Marquesans objects of compelling fascination for Westerners. The islands were named by the Spanish, claimed by the French — and then, in 1813, briefly occupied by the Americans in the person of Captain David Porter. Captain Porter, commander of the frigate *Essex*, had been instructed to challenge British shipping in the Atlantic during the War of 1812 but found himself inspired to attack British whalers in the Pacific. In this he succeeded well enough to ensure the later American dominance in Pacific whaling; and by the time it entered Taiohae Bay, Porter's *Essex* trailed in its wake five commandeered and armed British whalers. Captain Porter, who took up residence in the bay to rest his men and refit his makeshift flotilla, found Nukuheva an idyllic spot and the Marquesans inhabiting it ripe for the civilizing effects of colonial dominion. He declared the place an American possession, named it Madison's Island in honor of President Madison, and even plugged a written copy of his declaration into a bottle and buried it there. Porter, though, soon found himself embroiled in intercommunity antagonisms and ultimately concluded it necessary to invade the valley of the strongest and most ferocious people on the island, the Typees, and burn down their villages as a sort of grand civics lesson. He left Nukuheva two weeks after that shameful attack, never to return.

Porter's brief stay in the Marquesas marked the beginning and end of American political influence there. An American sloop of war, the *USS Vincennes*, did sail into Taiohae Bay in 1829, but the *Vincennes* is significant principally because she carried as a midshipman Herman Melville's cousin Thomas Melville and, as chaplain, a pious man named Charles Stewart. A party that included Stewart and Thomas Melville entered the Typee Valley, and the Reverend Stewart wrote about his experiences there in *A Visit to the South Seas* (1831) as part of an extended plea that the Marquesans, enduring "all the darkness of paganism," desperately needed the "enlightening and regenerating influences of Christianity" that might be provided by Protestant missionaries.[8]

By the time Herman Melville jumped ship in 1842, any American influence on the island, including a few physical remnants from an abortive missionary expedition, had faded, and the natives of Nukuheva were being pacified by five hundred French troops and seven French gunboats anchored in the harbor alongside the *Acushnet*. But the island was large and divided by precipitous ridges and thick vegetation; and to avoid capture and trial for desertion, the young Herman Melville and sailor Richard Tobias Greene quickly slipped out of Taiohae Bay and into the island's high and rugged interior. Coming out after a few difficult days and nights into another valley, the pair, injured, hungry, exhausted, fell into the care of a community of about two thousand islanders, the Typees, who had so far remained relatively isolated and culturally intact (in spite of Captain Porter's 1813 punitive invasion) because they were still regarded by other islanders and Europeans alike as ferocious warriors.

Herman Melville turned twenty-three during the three weeks he lived among the Typees in the latter part of July and early August of 1842.[9] He left or escaped on August 9 and was picked up by an Australian whaler, the *Lucy Ann*, which took him as far as Tahiti, where he was briefly jailed with some of the other crew on a questionable mutiny charge before escaping to another whaler. Surely three weeks was not much time to get to know the Typees, but it was long enough for the young Melville to acquire a basic background that, supplemented through reading other travelers' narratives, could be turned into his first book. *Typee* is an imaginative piece of literature in which the rough outlines of a real event have been passed through the mind of an artist. But Melville promoted his manuscript to publishers as a true story, absolute nonfiction. And when the book appeared at last in 1846, the author's preface boldly declared his tale to be strictly "the unvarnished truth."[10]

Typee is by no means the unvarnished truth, however, but more a search for answers to the problems raised for Westerners by the discovery of very different human societies in the East. *Where do we come from? What are we? Where are we going?* Paul Gauguin asked those three questions as he painted his great tableau. Melville and his predecessors on Nukuheva, gazing into humanity's mirror and considering the images presented by the exotic, elaborately tattooed, sexually expressive, warring and cannibalistic Marquesans, were forced to ask much the same questions — though framed a little differently. If the Marquesans represented humanity naked and in a precivilized state of nature, then the most obvious question they provoked was: Is humankind naturally evil?

Herman Melville structured *Typee* as a combination of adventure narrative, anthropological study, and political argument. His narrator, Tommo, is a man whose filtered vision shifts and flickers in the complex, anguished process of living among beautiful cannibals and trying to figure out what that exotic experience means. Tommo arrives in the Typee valley equipped to see things through practical and moralizing eyes, and his early anxieties about living among the notorious Typees surely parody the bluff arrogance and paranoid rigidity of earlier American and European commentators. He quickly rejects that limited perspective, however, and decides that the real savages in this world are not innocent South Seas islanders but aggressive Europeans: "The fiend-like skill we display in the invention of all manner of death-dealing weapons, the vindictiveness with which we carry on our wars, and the misery and desolation that follow in their train, are enough of themselves to distinguish the white civilized man as the most ferocious animal on the face of the earth."[11] Melville's Tommo begins to conclude that natural man (as typified by the Typees) is innately good and that therefore civilization, not the human heart, is the source of evil. Tommo makes the Typees' valley into a figurative paradise by punctuating his descriptions with overt references to the biblical Eden, and by stressing the beauty and physical perfection of the Typees, their happy innocence, artless simplicity, good-natured laziness, and the physical ease of their life. "The penalty of the Fall presses very lightly upon the valley of Typee," Tommo declares. No one has to garden or plant or hunt. Breadfruit and banana have always grown wild on the island, and a hungry appetite is easily satisfied by the casually outstretched hand.

No snakes. No predatory beasts. No mosquitoes. The South Seas setting of this book is indeed an almost perfect place, close to the biblical Eden. And Tommo uses that vision to suggest that the Western colonial powers, as they take over the island with their bristling gunboats and bible-thumping missionaries, are violating fundamentally a human peace in nature.

The image is powerful and appealing, but it leaves out a critical part of Typee life. In fact, as Tommo well knows, the peace of this Eden is punctuated regularly by intercommunity warfare. "Occasionally I noticed among the men the scars of wounds they had received in battle; sometimes, though very seldom, the loss of a finger, an eye, or an arm, attributable to the same cause."

How to portray a violent society as innocent? Tommo's first solution resembles Paul Gauguin's. Typee warfare was always conducted by the men, and Tommo adroitly focuses his narrative and descriptive powers on the young women. Indeed, the narrator insists he spent much of his time in the company of lovely adolescent girls, especially his own heart's favorite, a "beauteous nymph" named Fayaway. It is difficult now to reconstruct the actual truth about Typee when Melville was there. We do know that the author exaggerated his time spent in the valley fourfold. We know he overemphasized the ease of life in the valley by suggesting that no one had to labor when in fact the islanders, like other Polynesians, cultivated gardens and planted and tended their food trees. We also know that Melville shaped the geography of the valley to suit his narrator's imaginative needs. He invented, for example, a large lake in which to bathe and lounge on languid afternoons with his female companions, and on which, in one of the book's most memorable tableaux, the lovely Fayaway stood up in a canoe, spread out her loose robe to make a sail, and thereby revealed her naked and natural beauty.

There remained the problem of cannibalism. But only the Typee men ate human flesh; and they ate only the flesh of enemies killed in battle, which made it less reprehensible. Though they were cannibals, in other words, they were basically *nice* cannibals who, according to Tommo, strictly limited this otherwise appalling habit and were "in other respects humane and virtuous."[12]

Paul Gauguin's vision of the South Seas as a type of idyllic human past depended upon an absence of men. He painted Tahitian women and girls almost exclusively, presenting in portrait after portrait the knowing, artful image of a naive, artlessly sexualized Eve. Herman Melville's South Seas paradise also included a romantic place of the mind where lovely adolescent girls lolled around without many clothes on. But Melville's world was more complex, and in the end, the literary artist turned to his male warriors and cannibals and painted them in colors increasingly realistic — and threatening. After all his relaxed lounging, his playing and chatting and swimming with the "nymphs," Tommo, in the book's final few chapters, becomes increasingly concerned about who the Typee men really are, what they are doing — and what they are going to do. Once

Melville's Tommo turns to the men, which Gauguin never does, then the stain of evil starts to appear. First the narrator fears that the physically powerful warriors — themselves tattooed top to bottom, face to fingertip — intend to tattoo him. Then he discovers the ghoulish scraps of a cannibal feast, as well as three shrunken human heads wrapped up and kept in the house where he stays. This progressive discovery of evil in paradise, the violence and cannibalism among Typees, amounts to a realistic dissolution of the romantic vision and explains why, in the end, Melville's narrator chose to leave. He left this wonderful paradise, exemplified by the sexually free and innocent adolescent girls, because he was afraid of being beaten and eaten by the Typee men.

In the book's final scene, Tommo finds himself chased out of paradise by spear-throwing men and fleeing to a whaling boat that has entered Typee Bay to rescue him. Attacked at the last minute by a tomahawk-wielding Typee warrior, Tommo is compelled to smash him viciously with a boat hook. In short, Tommo's own violence mirrors the violence of these South Seas "savages" as he fights to reenter a Western civilization, which, for all its manifest crimes and corruptions, will still buy books about paradise in the South Seas.

If for his readers, the wonderful accounts of lovely, half-naked adolescent girls remained far more evocative than those final references to spear-throwing, tattooed warriors, for Melville himself clearly nature was not peaceful, and natural man was not a noble savage. It is doubly ironic that *Typee* remained for most people a true story, fundamentally an amateur anthropologist's study of Polynesia, until literary scholar Charles Robert Anderson demonstrated in his 1939 book, *Melville in the South Seas*, that the artist had falsified his chronology and borrowed from the writings of his predecessors.

For Herman Melville paradise was a piece of innocence that provided the perfect foil for his angry attacks against overdressed missionaries and scavenging colonialists. But the young

American writer was not fooled by the romantic abstraction of paradise with which he seduced his readers.

Preceding Herman Melville's *Acushnet* by little more than five years, a ten-gun brig of the British navy converted into a scientific survey ship, the HMS *Beagle*, rounded the horn of South America, moved in fits and starts up the coast as far north as Peru, and turned to wind slowly through the Galápagos Archipelago before catching the trades out into the deep South Pacific — past the Marquesas and bound for a ten-day layover in Tahiti. From there, the *Beagle* cut a wake into the Far East, across the Indian Ocean, around the African Cape, at long last dropping anchor in the English port of Falmouth on October 2, 1836, and disgorging, among many other homesick passengers, three live giant tortoises from the Galápagos Islands and one Charles Darwin.

"Tahiti is a most charming spot," the young Darwin wrote his friend and former teacher at Cambridge, the Reverend John Stevens Henslow. "Everything which former Navigators have written is true. . . . Delicious scenery, climate, manners of the people, are all in harmony."[13] Harmonious, indeed, but Darwin never imagined the island to be a paradise. However, Darwin's journey to the South Seas was indirectly responsible for others seeing a version of paradise in those islands because his *Origin of Species* would raise new questions about human potential and the possibility of social change. The challenges raised by Darwin's work meant that culturally isolated human communities, including those living on islands in the South Seas, would be revisited in the next century by anthropologists probing for newly sophisticated answers to Gauguin's questions. Among them, and the most influential of all, was anthropology's equivalent to Gauguin and Melville, a brilliant scientist whose conclusions suffered from being tailored to suit her preconceptions: Margaret Mead.

Darwin's *Origin of Species*, published in 1859, convinced most readers that the intricate workings of heritable biological processes were far more important to human existence, even to human culture, than previously thought. Exactly how biology and culture were related then became a vital question. In England, Darwin's cousin Francis Galton read *Origin* and decided with a burst of enthusiasm that "a great power was at hand wherewith man could transform his nature and destiny."[14] By 1874 Galton had plucked a Shakespearean phrase (from *The Tempest*) that turned the question of where we come from into a stark debate: *Nature* versus *Nurture*. And he, of course, supported nature.

The phrase was catchy, but a gross oversimplification. The reality is that all living organisms are influenced both by their genetic inheritance and by the environment they live in. It is true that in comparisons between two individuals we can often observe the influences of genetic or environmental differences. But only in comparisons can we do so, and even then only by the special trick of keeping either genes or environment constant. Both genes and environment influence hair color, hat size, and how we behave. Unrelated people with equally good nutrition, the same exposure to sun, and even the same hair dye can have differently colored hair. So genes — that is, nature — affects the traits. On the other hand, identical twin sisters can have differently colored hair because of variances in their nutrition, or unequal amounts of time spent in sunshine, or different choices of hair dye. So this second set of comparisons shows the importance of environment — nurture. Hair color is influenced by both nature and nurture, in other words. Those who look for the importance of genes would hold the environment constant and examine the results of a comparison. Those who want to find the influence of nurture, by contrast, would try to find a case where genes seem constant, and then look for differences imposed by experience. Each side can claim its victories, but to contrast those two forces in isolation from each other is absurd. So Galton's dilemma, nature or nurture, was a false one, an intellectual red herring. But it came to have such historical importance we feel it deserves a name of its own: Galton's Error.

Francis Galton felt he knew the answer to Gauguin's second question, *What are we?* We are creatures, he thought, directly arisen from nature, products dropped off the conveyor belt of a large Darwinian factory, the intellectual and moral consequence of nature, not nurture. This belief in a simplistic biological determinism he adopted early, and by 1883, after completing an extended study of twins, Galton insisted that he had achieved proof of "the vastly preponderating effects of nature over nurture." The publication in 1900 of Gregor Mendel's astonishing experiments with sweet peas showed how simple physical features of sweet peas could be genetically transmitted from one generation to another. Galton and his followers had little doubt that the more complex features of human behavior would sooner or later be shown to follow the same fundamental pattern, and by 1901 Darwin's cousin had initiated a grand crusade, a movement, he said, "like a missionary society with its missionaries" proceeding with "an enthusiasm to improve the race." The race he referred to was the human one, and Galton's plan for improving it — which he called *eugenics* — would follow the principles used in breeding domestic animals, that is, manipulating the reproduction of individuals to alter a group's gene pool.

Galton's views held alarming social implications, and it wasn't long before they were challenged. While Paul Gauguin painted away in Tahiti, while Gregor Mendel's experiments on sweet peas in an Austrian monastery were being resurrected, while Francis Galton prepared to establish his eugenics movement in Britain, in New York City a man who had read Kant sitting half-starved inside an igloo in the howling Arctic was granted tenure as a full professor of anthropology at Columbia University.

In 1900, with his first major lecture at Columbia, Professor Franz Boas began his lifelong campaign to challenge Galton's style of extreme biological determinism. Boas correctly perceived the danger that biological determinism could turn viciously racist; he correctly regarded culture as far more dynamic and powerful than the strict Darwinians of his time would have it; and Boas proclaimed his vision that anthropology should apply itself most forcefully to examining the mysteries of culture and its impact on human behavior. Pressed by the increasingly radical assertions from biological determinists of that period, by the 1920s he had declared himself actively searching for a way to distinguish experimentally between the biological and social origins of human behavior: to separate nature from nurture, claiming a "fundamental need," as he wrote in the *American Mercury* in 1924, for some "scientific and detailed investigation of hereditary and environmental conditions."[15]

One of his most eager and promising students at Columbia, Margaret Mead, was just then ready to begin her doctoral dissertation, and Boas decided Mead's dissertation ought to focus on adolescence. A demonstration, he felt, that coming of age was in one culture not stressful would indicate that adolescence as an emotional and behavioral entity was far more a product of nurture than of nature. A negative instance would destroy the claim of universality and sway the nature-nurture debate back toward nurture. "Are the disturbances which vex our adolescents," Mead asked rhetorically, referring to American and European teenagers, "due to the nature of adolescence itself or to the civilisation?"[16]

To answer that question, Mead followed Melville and Gauguin to the South Pacific. On August 31, 1925, at the age of twenty-three, the young American walked down the gangplank of a Matson cruise ship into the exotic port of Pago Pago, on the island of Tutuila in American Samoa. Her findings from this expedition would capture the imagination of the Western world and galvanize a movement toward cultural relativism. Yet she was later proven extraordinarily wrong in many of her claims about Samoan life.

Samoa is an archipelago of volcanically formed South Seas islands, nine of them inhabited by a single cultural group who call themselves Samoans and speak the Samoan language. Politically, the islands are currently divided between Western

Samoa (four islands, independent since 1962) and American Samoa. Mead would regularly emphasize the remoteness and cultural primitiveness of the islands. But when she arrived in Pago Pago, she entered a Polynesian society that had been Christianized by Protestant missionaries some eighty years before and had been a legal territory of the United States for more than twenty years. Her disembarkation was accompanied by the sounds of the United States navy band and the sights of several American battleships and airplanes; she carried a letter of introduction to the Surgeon General of the American navy and soon enough was given the honor of dining with the admiral of the Pacific fleet.[17]

Mead was later to recall that "through the nine months" she spent in Samoa, she "gathered many detailed facts" about "all the girls of three little villages" on the remote island of Ta'u. "Speaking their language, eating their food, sitting barefoot and cross-legged upon the pebbly floor, I did my best to minimise the differences between us."[18]

Her readers could easily imagine that she spent nine months, day and night, living in primitive conditions directly with her subjects — but that would be an exaggeration almost as extreme as Herman Melville's claim to have spent four months among the Typees. Margaret Mead actually spent around six months on the island of Ta'u, with approximately three of them devoted to interviewing the girls. Since she arrived in Samoa completely unfamiliar with the Samoan language, she found it necessary to stay in her hotel room in Pago Pago on the main island of Tutuila for at least six weeks while a language teacher came once a day for an hour's lesson.[19] Those brief lessons were inadequate, and, as a further attempt to learn the language, for ten days Mead tried living in a Samoan household not far from Pago Pago. By late October, though, she had decided to leave Tutuila altogether and begin her field work on the more remote Ta'u; on November 9, she was given a lift to that island on a U.S. navy minesweeper. Mead's ten days in a Samoan household on Tutuila had been time enough to convince her that she preferred to live in the Western style. She was very reluctant to stay with Samoans, so she wrote home to Boas, because she feared a "loss of efficiency due to the food and the nervewracking conditions of living with half a dozen people in the same room in a house without walls, always sitting on the floor and sleeping in the constant expectation of having a pig or chicken thrust itself upon one's notice." The only non-Samoan household on Ta'u was located in the U.S. naval pharmaceutical dispensary, where the navy pharmacist, Edward Holt, and his wife and children lived. Mead chose to live with the Holts, who gave her a private room and permission to use a small house in front of the dispensary for conducting her interviews and tests.

She began her research. From three villages on the island, Mead studied in detail fifty girls and young women, including twenty-five who had not begun menstruation and twenty-five who had. The twenty-five adolescents, between fourteen and twenty years old, were her central study group. Her interviews and tests proceeded from mid-November to early March — with significant interruptions caused by a devastating hurricane in January; the arrival of a European shell-collecting expedition in mid-February; and the resumption of mission schooling in late February.

Mead was later to make many sweeping claims about Samoan culture in general, based largely on an expertise acquired during her single nine-month stay — arguing that although no one could be expected to become an expert on complicated European societies in such a short time, Samoan culture was in fact very simple, and "a trained student can master the fundamental structure of a primitive society in a few months."[20] That assertion seems, from a modern perspective, presumptuous. It becomes more so when we realize that during Mead's sojourn on Ta'u, all political, economic, religious, and ceremonial decisions were made by councils of men only. She was thus excluded from observing at firsthand many significant aspects of Samoan cultural life. She was forced to rely almost entirely on her adolescent girls for most of her direct information about the wider culture.

Mead left Ta'u in May of 1926 and began her long voyage home. She returned to New York, won a position as assistant curator at the American Museum of Natural History, and within a year, by the spring of 1927, had virtually completed the typescript for her book, *Coming of Age in Samoa: A Psychological Study of Primitive Youth for Western Civilisation.*

Coming of Age, which appeared in 1928, argued a fundamentally simple thesis: Nurture far more than nature wrote the human script. Societies could choose to construct extraordinarily different sets and sequences of behaviors for people to act out almost as freely as individuals can choose the clothes they wear. The proof of this thesis was that Mead had investigated adolescence in the South Seas and found it to be amazingly different from adolescence in the West. While coming of age in the West was a time of "stress and strain,"[21] in Samoa the same developmental phase was for a girl "the best period of her life." Thus, American-style adolescence was not a universal and inevitable consequence of biologically driven feelings, passions, and behaviors, but rather an unnecessarily painful production of a sexually repressive Western culture.

Growing up in Samoa was "so easy, so simple a matter," Mead wrote in *Coming of Age,* partly because of the "general casualness of the whole society,"[22] but mostly because sex was regarded as "a natural, pleasurable thing." In such a relaxed culture, uncorrupted by the repressive influences of Western Protestantism, there was "no room for guilt." Just as Samoan children were raised in a warm but emotionally undemanding and fundamentally permissive style, so the pubescent girl found no restrictions on her inclinations to pursue a broad variety of sexual partners. Adolescence was a wonderful period of free and open sexuality, a time of delightful, lighthearted promiscuity.

Given how successfully the Samoan culture had constructed its sexual attitudes and behaviors, moreover, most psycholo-

gical problems, typical of the "maladjustment which our civilisation has produced,"[23] simply vanished. Jealousy was rarely
aroused, for instance. Samoans rarely developed neuroses or
marital problems or Oedipus or Electra complexes. As a matter
of fact, no Samoan woman was ever frigid and no Samoan man
ever became impotent from psychological causes. And, with no
particular reason to be unhappy, there was, Mead eventually
concluded, virtually no suicide.

In this apparent paradise, the anthropologist continued to
inform her readers, the culture had chosen such a satisfactory
attitude toward sexuality that rape just about disappeared; adolescent boys were too busy making sure the voracious girls were
sexually contented. In *Coming of Age*, Mead was actually somewhat circumspect on the issue, acknowledging that "rape, in the
form of violent assault," did occur "occasionally" in Samoa —
but it was surely the fault of "contact with white civilization."[24]
In some of her other writings on the subject, Mead became more
certain, asserting at one point that "the idea of forceful rape or of
any sexual act to which both participants do not give themselves freely is completely foreign to the Samoan mind." Of
course there was, as she candidly admitted in *Coming of Age*, a
"peculiar abuse" perpetrated by the *moetotolo*, or sleep crawler.
The sleep crawler was a boy or man who would sneak into an
adolescent girl's bed when she was expecting her lover in order
to trick the girl into having sex with him, to "stealthily" appropriate "favours which are meant for another." In the West, such
an act might be regarded as simply one form of rape, but in
Samoa, so Mead declared, a sleep crawler merely "complicates
and adds zest to the surreptitious love-making." Catching one of
these sneaky fellows in the act was "great sport."

Not only had the culture simply eliminated adolescent
angst, parental repression, all neuroses, most jealousy, all frigid-
ity and impotence, and most rape, it had also done away with
violence: "No implacable gods, swift to anger and strong to punish, disturb the even tenor of their days."[25] What warfare there
was in Samoa, she wrote elsewhere, had been stylized, merely a

consequence of village squabbles and therefore killing only one or two unlucky people at a time. In fact, Samoans would "never hate enough to want to kill anyone." They are among the "most amiable, least contentious, and most peaceful peoples in the world."

Coming of Age became almost immediately a huge popular success. Many readers, no doubt, were stimulated by Mead's imaginative and mildly titillating word pictures of a free-loving South Seas paradise (or by the bare-breasted beauty and her lover rushing across a beach under a full moon, as featured on the original dust jacket illustration).[26] But Franz Boas and his associates and colleagues, students and former students, endeavored to tilt the reception of the book to more elevated levels. A foreword by Boas declared the study to be a "painstaking investigation" confirming "the suspicion long held by anthropologists, that much of what we ascribe to human nature is no more than a reaction to the restraints put upon us by our civilisation."[27]

Mead herself never returned to Samoa and never altered the text of *Coming of Age*. In a 1961 introduction to the book, she compared her portrait of adolescent girls in the South Seas to the eternal lovers who stand forever as perfect art on the glazed surface of Keats's Grecian urn. Her painting of the Samoan paradise would, she declared, "stand forever true because no truer picture could be made of that which is gone."[28]

But any number of European explorers, traders, adventurers, missionaries, and government officials had been coming to the islands since 1722. Their written reports regularly contradicted Mead's, as did the observations of several social scientists who came to parts of Samoa subsequently.[29] An Australian anthropologist named Derek Freeman, who began his own field work in the archipelago fifteen years after Mead, and who between 1940 and 1981 spent a total of six years living intimately among Samoans, finally published in 1983 the first full analysis and critique of Mead's work, *Margaret Mead and Samoa: The Making and Unmaking of an Anthropological Myth*. Freeman's analysis has been characterized, correctly, as a "frontal attack."[30]

Indeed, the tone of this book is ultimately unforgiving and stridently polemical;[31] and it suffers from an overly simplistic analysis of the intellectual context of Mead's work. It oversimplifies Boas's stance as an anthropologist, for example, and it appears to exaggerate Mead's influence on the thinking and methodology of subsequent cultural anthropologists.[32] But Freeman's more particular assertion, that Mead greatly overgeneralized from a limited set of data, looks correct.

As Freeman reminds us, Margaret Mead did not actually study adolescence in the United States, nor did she review what was known scientifically about adolescence in the West, and so her comparison of the two cultures is weakened by the fact that she provides data for only one. Though Mead never actually studied boys in Samoa, she nonetheless began generalizing, by the 1930s, about the nature of adolescence for them as well. In addition, her study group was far from ideal. Her twenty-five adolescent girls included three who were, according to Mead's own assessment, "deviants" in the sense that they "rejected the traditional choices."[33] In addition to those three, another three were by Mead's description "delinquent." But the fact that six of her group of twenty-five were maladjusted was essentially forgotten.

Adolescence was for both sexes the "age of maximum ease" in Samoa, so Mead wrote in 1937. But in fact, so Freeman points out, police records show the delinquency rates for adolescents in Samoa to be consonant with those of other countries; and the ratio of male to female first offenders in Western Samoa parallels the typical ratio for Western countries: five to one.[34]

As for the remarkably carefree promiscuity of Samoan girls, Mead's own account indicated that fourteen of the twenty-five pubescent girls in her sample — distinctly over half — were virgins.[35] Indeed, the Samoan people are distinguished from some other South Pacific island cultures by having a traditional obsession with virginity, as expressed particularly in the institution of the *taupou*, or ceremonial virgin. Virgins were and are highly valued in Samoa, and thus, reports Freeman, the activities of an adolescent girl are carefully observed by her brothers, who, should they find her in the company of a potential lover, are likely to berate her, possibly beat her, and assault the boy.

Mead described jealousy as a rare emotion in Samoa, and adultery often not the source of much "fuss." In fact, several historical accounts of Samoa describe jealousy as a frequent and severe emotion and note that the ordinary punishment for adultery was death. The deceived husband was free, by Samoan tradition, to seek revenge on any member of the guilty male's family, while the adulterous wife would likely be punished by the severance of a nose or ear, or the breaking of bones.[36]

Mead insisted that the occasionally occurring violent rape in Samoa was the result of contact with Western civilization, and she distinguished between forcible rape and the traditional sleep crawling. In fact, both styles of rape were and are common in Samoa. Historical accounts as far back as 1845 describe instances of forcible rape; the first court records for American Samoa, starting in 1900, frequently detail rapes. In the 1920s, when Mead visited the islands, the *Samoa Times* regularly reported rape cases; and jail statistics from the period indicate that rape was the third most common crime on the islands. In the 1950s, government statistics reported rape to be the fifth most common crime. Many forcible rapes in Samoa were adjudicated within the local community, but the cases officially recorded by Samoan police during the late 1960s suggest a rate twice that of the United States and twenty times that of England for the same period.

Mead's generalizations about the peacefulness of Samoan society — no war gods, no wars, little serious contention or hatred or violence, and so on — are all, according to a wealth of historical, anthropological, and contemporary information, wrong.[37] Half of the seventy-some main gods in pre-Christian Samoa were war gods. By all accounts, wars in pre-Christian Samoa were common and very bloody. John Williams, a missionary and explorer who visited the islands during the 1830s, observed an eight-month-long war between two regions of Samoa and described regular battles between hundreds of participants. The victors in this war tore out the hearts of some of their captured enemies; four hundred other captives, including women and children, were burned alive. Williams arrived on the island of Ta'u — where Mead did her study — in 1832 and learned that a major war between Ta'u and the neighboring island Olosega had taken place four months earlier, during which thirty-five men from Ta'u, more than a tenth of the total number of adult men on the island, lost their lives. Indeed, warfare between Ta'u and Olosega was so persistent that a sporadic series of raids and retaliations continued even past the time of Mead's visit.

Though Mead declared the Samoans to be among the "most amiable, least contentious, and most peaceful" of peoples, police records for 1964 to 1966 tell us that Western Samoa had five times the rate of common assault as occurred in the United States during the same period. Police records also indicate Samoan rates of serious assault (assault causing physical injury) to be more than half again the U.S. rate for the same period, almost five times the Australian rate, and eight and a half times the New Zealand rate. Though Mead insisted that Samoans "never hate enough to want to kill anyone," police records and other reliable sources sketch a completely different picture. Western Samoa's murder rate for 1977 was almost half again as high as the U.S. rate during a comparable period, while American Samoa's murder rate was about five and a half times that of the United States.

Margaret Mead was a bold pioneer and a gifted writer with a special talent for bringing academic insights into the public arena. The fundamental lessons of her early work in Samoa — that Western ideas of human possibility were limited, that Western sexuality was too inhibited, that child rearing in the West was rigid and overly authoritarian — were welcomed and taken to heart by the culture at large. A mother herself, Mead adopted breast-feeding on demand after observing the practice during field work in New Guinea; and so she was able to persuade her pediatrician, Dr. Benjamin Spock, that on-demand feedings would produce happier, healthier children. That message, too, was thoroughly passed on.

Coming of Age in Samoa became a classic, an essential text for introductory courses in the social sciences, the most widely read piece of anthropology in history, and it made its author famous beyond what anyone could have predicted. She spent her life promoting it and the ideas it represented, simultaneously witnessing and encouraging her own apotheosis. Ultimately, she grew into "a symbol of all anthropology," and found herself transmogrified (according to *Time* magazine in 1969) into "Mother to the World." For Mead herself, paradise *was* Samoa. While Samoa, in turn, became for the general public pristine evidence that culture alone — nurture without nature — scribes its mysterious markings on the blank slate of human character.[38] The shortcomings of Mead's Samoan research are no more remarkable than those of any number of dissertation projects past and present; but its astonishing success and its unwitting transformation of a pastoral fantasy into ultimate proof for an extreme position of cultural determinism helped perpetuate for another half century the misleading separation of nurture from nature first suggested by Francis Galton in 1874.

Cultural determinism, as the counter to Galton's biological determinism, leads us to hope and to move practically in a difficult world; but it can also bring us to oversimplify necessarily complex problems and to avoid examining hard realities. It can lead to denial, and the regressive creation of a mythical Arcadia, a golden age, a paradise in the remote tropics, or a perfect time and place somewhere else where most human problems are solved by easy choice and a few basic, often tax-free decisions. As Newt Gingrich, current speaker of the U.S. House of Representatives, expresses the concept: "We had long periods in American history where people didn't get raped, people didn't get murdered, people weren't mugged routinely." Such crimes, the speaker concludes, are therefore entirely "social artifacts of bad policy."[39] Well, not entirely. Even with good policy they are hard to eliminate. We would be foolish to think otherwise.

And what of the real Arcadia in ancient Greece? "Et in Arcadia ego" — I too am in Arcadia — reads the inscription on an anonymous tomb, a sentiment that inspired paintings of heaven by Poussin. Its mountains and fertile valleys had been an inspiration, but in real life Arcadia could not remain the haven from foreign dominance that it was in 1100 B.C. By 500 B.C. Arcadia had joined a military alliance with Sparta in the Peloponnesian League, so that even that vision of paradise fell to the reality of violence. "Et in Arcadia ego"? Yes, we are all in Arcadia, but Arcadia is not paradise. It is a place where wars and other evils continue to threaten and occasionally to happen. There is no such thing as paradise, not in the South Seas, not in southern Greece, not anywhere. There never has been. To find a better world we must look not to a romanticized and dishonest dream forever receding into the primitive past, but to a future that rests on a proper understanding of ourselves.

6

′ ′ ′

A QUESTION OF
TEMPERAMENT

I T'S ALL VERY WELL to conclude that neither in history nor around the globe today is there evidence of a truly peaceful society. But the suggestion that chimpanzees and humans have similar patterns of violence rests on more than the claim of universal human violence. It depends on something more specific — the idea that men in particular are systematically violent. Violent by temperament.* This notion of the violent male seems reasonable to anyone familiar with crime statistics, and

---

* Temperament, as we define it, is the emotional element of personality. It is a system of emotional systems; in other words, temperament consists of an individual's emotional reactions to situations in the real world. The idea also includes predictability. Individuals may have predictable reactions to a given set of circumstances. Temperament varies between individuals, but it is also possible to think about the general or average temperament of a sex. Likewise, it's possible to talk about the temperament of a species, in the sense of an average of individual temperaments in that species. Our use of this concept partially follows the work of Susan Clarke and Sue Boinski (1995), though we apply it more broadly.

explains why we can't find paradise on earth. Even so, it's hard to shake the nagging suspicion that our own Western societies bias our perspective. Perhaps there is some truth to the idea of an Amazon nation, where violence is the special preserve of women. And what about those hunter-gatherers? Even if they are sometimes violent, doesn't their egalitarian ethic nevertheless challenge the concept of the aggressively dominating male?

There's a more theoretical level, too, where we meet the same sort of resistance to characterizing men as violent. This is the widespread belief that "gender" is culturally determined, an idea difficult to shake partly because it has been a bulwark in traditional feminist struggles for equality. Almost everyone agrees that male violence is commonplace in the West. So is the violence of males merely a Western invention? Perhaps we can find counterexamples, places where women are by cultural license or tradition violent. Maybe there is a society of Amazons somewhere in the world, with women warriors, or at least with women dominant and running the show.

The most extreme example of women warriors on record is the case of the Dahomey elite guard. This force of "Amazons," as admiring European commentators described them, constituted an all-female army in the Dahomey kingdom of West Africa (now the Republic of Benin) during much of the nineteenth century. They reached their high point in 1851 during the bloody rule of King Gezo, when the women's force was expanded and trained to serve in active military campaigns, supplementing with perhaps 5,000 fighters a total army that may not have been larger than 12,000 altogether.[1] Europeans described the Amazons conquering parts of the country, carrying scalps and taking captives. But their military strength never recovered from a massive defeat by the neighboring Abeokutans in 1851, when as many as 2,000 Amazons were reported killed. Still, their reputation was impressive. The explorer Richard Burton said, "The women were as brave, if not braver than, their brethren in arms."[2]

Yet it is not clear whether belonging to that force represented a liberation for women or simply a more brutal than usual form of exploitation. The women's force may have begun in the eighteenth century as a contingent of the king's wives, armed and serving as a largely ceremonial palace guard, necessary since Dahomey kings banned all men but themselves from the palace. In some ways, then, the guard were little more than fighting members of the king's harem: compulsorily recruited from the palace or through capture, and sometimes by conscription, but in any case always regarded as the king's property. According to one tradition, upon recruitment the Dahomey women were forced to submit to clitoridectomy, on the theory that it would reduce their lust. And since the king had sex with only a select few of these women, any pregnancy among the others was evidence of adultery, punishable by death. The Dahomey women warriors themselves, ironically enough, seemed to embrace their own culture's more traditional stereotypes about gender by declaring that in becoming fierce soldiers they were reborn as men. As Tata Ajachè, a former Dahomey warrior, recalled in later years, after she had disemboweled her first victim in combat, she was embraced into the sisterhood by being told: "You are a man!"[3]

Most societies around the world restrict women entirely from participating in war and usually even from involvement in planning or discussing it. This was the case in fifty-eight out of sixty-seven societies in the ethnographic record surveyed for female warriors; and in the nine societies with some women's participation in war, they were always less involved than men. Navaho women, if they wished, could join a war party. They fought "just as did the men," but there were never more than two women at a time in a raid. Maori women joined the fighting on occasion, and Majuro women of the Marshall Islands, "although in the minority," would fight by throwing stones during a battle. Women of the Orokaiva were "always ready to urge on the fighting men and even to mingle in the fray." The Crow claim memories of one woman who went to war. Fox women were allowed to participate in war; there were "even some women who have become warrior women." Delaware women seldom went to war, but they had the right to if they chose. Comanche women would "snipe with bows and arrows on the fringes of the fray."[4]

Among traditional peoples using primitive weapons, it is easy to explain why there were so few women warriors. Men are on average more than four and a half inches taller than women and carry on a frame of denser bones a greater ratio of muscle to fat tissue.[5] Modern weapons, usually powered by a chemical explosion, tend to eliminate the significance of human physical differences, but traditional weapons have the reverse effect. Since they depend largely upon leverage to increase lethal penetration, traditional weapons actually exaggerate original differences in upper body strength — but even the original differences are hard to ignore. So women warriors were rare at least in part because, presumably, they were less effective than men.

But with modern weapons, of course, the situation is different, and women have demonstrated their fighting worth within the military organizations of many industrial societies — most often during defensive crises where the survival of the larger community seemed at stake. Germany during World War II promoted an extreme sexual conservatism and so excluded women from all significant military positions until the very end when, as their enemies closed in on all sides, a desperate Nazi high command drew up tentative plans for a women's combat battalion.[6] The Soviet Union, which regularly pushed a Marxist ideal of full sexual equality, sent many women directly into combat soon after their early losses in World War II; and the Soviet case represents the most complete modern attempt to integrate women into warfare. All Soviet women without children and not already engaged in critical industries were called up after the German invasion and occupation of 1941, and they thus became machine gunners, infantry soldiers, and snipers, as well as specialists in communication and transportation. They marched, shot, drove tanks, piloted fighters and bombers, and became particularly concentrated in antiaircraft units. It was not uncommon for women to serve in the same units with their husbands; and when Mariya Oktyabrskaya's husband, a tank commander, was killed, she got her own tank, christened the *Front-line Female Comrade,* which she pressed into battle until her own death. Altogether, roughly 8 percent of Soviet combatants during the war were women, and after the war they were given 4 percent of the total military decorations (but less than 1 percent of the nation's highest award, the Hero of the Soviet Union medal).[7] Once the national crisis was over, though, women in the Soviet military were demobilized; the peacetime draft was directed at men only. Women were still allowed to volunteer, but they came into an army that trained them less intensively than men, gave them special regulations, lower expectations, gentler punishments, narrower duties, and fewer promotions than men. The Soviet postwar army was all male, except for some specialized female support.

During the seven-and-a-half-year Algerian war of independence against the French, a struggle conducted largely through urban terrorism, many women were said to have discarded their veils and fought right beside the men. The reality is that the Algerian women served largely in supporting roles in that effort and were only involved in about 2 percent of the instances of actual violence, such as the planting and triggering of a bomb.[8] During the 1947 to 1949 Israeli war of independence, 114 of the some 4,000 Jewish fighters killed, or nearly 3 percent, were women. Upon the formation of the state of Israel, however, even though women became subject to a draft, the terms of conscription excluded women from combat. Women soldiers learned how to drive tanks and even became instructors in tank warfare, but they were not permitted to point their machine in the direction of a fight.[9]

So women have fought alongside men in wars, particularly in wars of defense against an alien occupation. Yet even for modern nations socially or ideologically committed to sexual equality, and even when the nation's existence has been seriously threatened, men have done the great majority of the killing. Yes, indeed, there are women warriors. But always they are women serving in a men's army, fighting men's wars.

Turning away from warfare between communities and toward the matter of violent crime within a community, we find much the same gender-biased pattern. Women have murdered, of course, and there have been women bank robbers, terrorists, kidnappers, the rare gangster, and an occasional woman rapist (typically of other women). There are most certainly female criminals of every stripe and spot, but everywhere, as a globally consistent trend, the gender of the criminal population correlates predictably with the violence of the crime. Male criminals specialize in violent crime. In the U.S., for example, a man is about nine times as likely as a woman to commit murder, seventy-eight times as likely to commit forcible rape, ten times as likely to commit armed robbery, and almost six and a half times as likely to commit aggravated assault. Altogether, American men are almost eight times as likely as women to commit violent crime.[10]

Even when we consider nonviolent crime, the gender factor is equally strong. American men are nearly thirteen and a half times as likely to commit fraud, thirteen times as likely to be arrested for carrying or possessing a weapon, more than ten times as likely to burgle, nine times as likely to steal a car, eight and a half times as likely to find themselves collared for drunkenness, and well over eight times as likely to be pinched for vagrancy. They are eight times as likely to vandalize, nearly seven and a half times as likely to fence stolen property, seven times as likely to commit arson, six and a half times as likely to be arrested for gambling offenses, six and a half times as likely to be stopped for drunk driving, and some five and a half times as likely to be hauled in for sex offenses (excluding prostitution and forcible rape). And they are five times as likely to be taken in for drug abuse offenses, four and a half times as likely for offenses against children and the family, over twice as likely for larceny, almost twice as likely for forgery or counterfeiting, and one and a half times as likely for embezzlement.

In only two criminal categories do American females lead males. First, adolescent girls are a little more commonly arrested for running away from home. Second, women are almost twice as often arrested for prostitution and other forms of "commercialized vice."

Is the overwhelming proportion of male violent crime in the United States just one more evidence of the patriarchal structure of American society, likely to be equaled by female violence as soon as women achieve equal status and power? It is certainly true that as overall crime rates in America have increased, so have the number of crimes committed by women. At one time, feminists Freda Adler and Rita Simon promoted the concept of an "explosion" in female crime, attributable to a positive liberation in roles and self-image created by the women's movement. Other observers, including Naomi Wolf and Liz Weil, have reported on the appearance of a new, sexually aggressive and stylistically masculinized (combat boots, dirty talk, cigars) "bad girl" — based on the concept that "If you can break free from the culture's ideas about sexuality, you can break free from anything."[11] But, as Wolf and Weil will be the first to acknowledge, the "bad girl" style is ultimately a rebellion from constraining expectations, not a promotion of violent behavior. In spite of media attention to the still newsworthy occurrence of violent crimes committed by women,[12] statistics do not support any fashionable concept of the new woman criminal.[13]

Is the frequency of male violence merely an artifact of male strength? When women become as strong as men, in other words, will they be inclined to be as violent? A good way to answer these questions using real-life data is to examine crime statistics from around the world in which murderer and victim are the same sex — murders for which, in other words, neither victim nor victimizer has an advantage based upon a male physique. What we find from these statistics, gathered from three dozen human communities around the world, is utterly clear and amazingly consistent. Crime statistics from Australia, Botswana, Brazil, Canada, Denmark, England and Wales, Germany, Iceland, India, Kenya, Mexico, Nigeria, Scotland, Uganda, a dozen different locations in the United States, and Zaïre, as well as from thirteenth- and fourteenth-century England and nineteenth-century America — from hunter-gatherer communities, tribal societies, and medieval and modern nation-states — all uncover the same fundamental pattern. In all these societies, with a single exception, the probability that a same-sex murder has been committed by a man, not a woman, ranges from 92 percent to 100 percent.*

Feminists have intensively debated the issue of *difference:* whether men and women really are distinctively different in behavior and temperament, whether focusing on gender differences is politically useful or regressive, and from whence the differences may have arisen.[15] But for reasons practical and historical, traditional feminists have stopped short in their analysis, maintaining a cultural-determinist outlook on gender in the style and tradition of Margaret Mead.[16] Cultural determinism sees all important gender differences as invented by and transmitted through culture — society. And, though feminist theorists have developed some provocative anthropological and psychoanalytical explanations to account for the transmission of difference, the cultural theories of gender look most compelling when expanded to fit a historical framework: the vision of patriarchy. According to patriarchy theory, gender difference is an artifact of particular historical events. Men have become the

---

* The one exception, Denmark, registers a probability of 85 percent, only moderately below the rest. If we eliminate the special situation of infanticide and focus only on adult men killing other adult men versus adult women killing other adult women, then the ratio of male to female grows even higher. In Denmark, for example, 100 percent of same-sex-adult murders were committed by men.[14]

way they are — in a word, patriarchal — because they live in a patriarchal society. Women have become the way they are for the same reason. That is to say, men are enlarged and validated by living in a patriarchal world, whereas women are thwarted and distorted by the same cultural experience. The idea may be ultimately circular, perhaps, yet it is compelling if only because we in the West do indeed live within a patriarchal tradition, where an inequality of the sexes in favor of men is institutionalized at all levels.

The most simplistic version of patriarchy theory presumes that male dominance is the particular creation of Western civilization, a cultural crime begun at some point in preclassical times and since then perpetrated by white males. British historian Roy Porter argues that sexual violence — rape — is directly associated with the Judeo-Christian patriarchy. "Feminists have pinpointed sexual violence in Western society," he has written, "and exposed its foundations in ideology and power relations. Of all world civilizations, the West is unique in its unrivalled powers of conquest by military might and colonial expansion. Trade follows the flag, and capitalist economics have likewise drawn on quasi-military goals of competition, expansion, struggle, victory, under the leadership of 'captains of industry.'" Western male-dominated culture is fundamentally violent, according to Porter, and this ideology of violence has been turned outward, in the militarized domination of other societies, and inward, in a parallel militarization against women. "The Western mind thus possesses a vast cultural reservoir of phallocentric aggression directed against women."[17]

But people like Porter are simply mistaken in limiting such violence to the West. Though it is clear that different societies carry, as part of their cultural codes, various attitudes toward self and others and various degrees of tolerance for violent behavior, the West's cultural and historical sins are not unique or even very distinctive. The world's most populous nation, modern China, exists as a direct consequence of patriarchal empire-building that began as early as the Ch'in dynasty, two centuries before the birth of Christ. A patriarchal, militaristic society in Japan pursued dominance and empire in Asia during the middle of our own century, and while they were still winning, during the month of December in 1937, the Japanese army paused to rape 20,000 women in the former Chinese capital city of Nanking.[18] In 1971, the all-male army of Pakistan, in the process of attempting to keep East Pakistan from becoming Bangladesh, killed perhaps 3 million people and raped between 200,000 and 400,000 unarmed Bengali women.[19] Polynesian expansion and conquest during the last 2 millennia across a vast portion of the Pacific Ocean established on 38 major islands and archipelagos an entire constellation of societies centered on the authority of the king and enforced with the power of the *toa* — a word that means both warrior and ironwood tree, the source of Polynesian war clubs and other weapons.[20] And Shaka, patriarch of the Zulus in nineteenth-century southern Africa, having organized a professional fighting force and developed the stabbing *assegai* for disemboweling enemies in close combat, created a great empire through "battles of annihilation" and in the process effected a diaspora of refugees across one-fifth the continent.[21]

Bartolomé de Las Casas, a Dominican bishop living in the New World, observed at firsthand many of the atrocities committed there by the Spanish and wrote about them in his *The Devastation of the Indies: A Brief Account* (1542).[22] The conquistadors, sometimes assisted and sometimes resisted by their own Christian counselors, killed tens if not hundreds of millions, brutally plundered and then erased the civilizations of the Aztecs in Mexico, the Mayas in Yucatán, and the Incas in Peru. No one should imagine it is in the Spaniards' defense to point out that the civilizations they destroyed and displaced were themselves colonialist empires, somewhat equivalently patriarchal, rapacious, and cruel.[23] The Aztecs in particular fought regular battles with neighboring peoples both for territorial consolidation and, more simply, to provide a steady supply of sacrificial victims to propitiate their insatiable gods. After an unsuccessful revolt by one colonized people, the Huaxtecs, their Aztec masters took 20,000 captives for the dedication of a new pyramid temple in the Aztec capital. On dedication day the Huaxtec victims climbed the steps to the truncated top of the pyramid, whereupon their chests were pried open with an obsidian knife and their still-pumping hearts ripped out and held aloft.

The violence, cruelty, and destructiveness of Western societies are clear. But such vices are not so distinctively Western. Islam, to consider yet another alternative, looks approximately as patriarchal, bellicose, and expansionist as Christianity. The great prophet and patriarch, Muhammed, himself was a warrior and, unlike Christ, called for spreading the faith by sword: the *jihad*, or holy war, to be advanced against anyone defying the will of Allah. The Muslim concept of the world was thus divided into two pieces, the Dar al-Islam, or House of Submission, and the Dar al-Harb, or House of War. By the seventh century A.D., Islamic warriors had expanded their House of Submission outward into the contemporary lands of Arabia, Syria, Iraq, Egypt, and North Africa. By the middle of the ninth century, patriarchal Muslim states throughout the Mediterranean and Middle East were retaining their shape as best they could with the help of a slave-warrior class largely acquired from Turkish peoples to the East.[24]

The West acquired the secret of gunpowder from the Chinese and combined it with a native tradition of metallurgy to create Europe's arsenal of weapons with hugely superior penetrating power. With the best weapons and navigational technologies, Europeans pursued global capitalism and colonialism. But Western males did not invent violence against men, violence against women, warfare, empire building, or patriarchy. And imagining the victims of European colonialism as a sorry series of feminized — or nonpatriarchical — gentle societies is simplistic, patronizing, and wrong. A 1971 survey of ninety-three societies around the world found men to hold the bulk of the political power in all of them and to retain *all* important political positions for 88 percent of them. Outside the more formalized and public political systems, men in 84 percent of the societies held *all* the significant leadership roles in kin groups as well.[25]

Any full analysis of patriarchy should take into account the veiling, sequestering, and regimentation of women in Muslim societies; the tradition of foot binding in China; the suttee tradition of the Indian subcontinent; the deeply institutionalized practice of clitoridectomy among many cultures in 26 different nations across the African continent, a mutilation affecting 2 million girls each year;[26] the near ubiquitousness of wife beating around the world;[27] and the fact that polygyny — multiple wives — is an accepted practice in far more cultures than not. It ought to examine the fact that 67 percent of married women in rural Papua New Guinea describe themselves as battered, with one in five hurt severely enough to require hospitalization at least once.[28] It should note that in Pakistan the mortality rate for girls is half again as high as that for boys, who are better fed; and that in many parts of Africa and the Middle East women cannot visit health clinics without their husbands' permission.[29] We might hear more about the problem of patriarchy outside the West if more non-Western women were freer from grinding poverty and, in some instances, rigid cultural restrictions against expressing themselves.[30] In Bangladesh, one must remember, feminist poet Taslima Nasrin has been placed under *fatwah*, the Muslim death sentence, for writing the wrong thing.[31]

If patriarchy is purely a cultural construction, an arbitrary occurrence in history, then logically we should be able to find places and times where it does not exist. Johann Jakob Bachofen, a German attorney, presented in 1861 his influential theory (in *Das Mutterrecht — The Mother Right*) that women, through their maternal presence, took humanity out of dark barbarism and instituted the start of culture and civilization, a new social system that was at first matriarchal.[32] Bachofen's theory of a matriarchal stage in human history is weakened by the fact that no anthropologist has yet to find anywhere a genuine matriarchy, a society where women actually rule in a system that mirrors patriarchy. Nor has anyone discovered convincing archae-ological evidence for the existence of a matriarchy in the past — in that, theorists from Margaret Mead to Helen Fisher to Gerda Lerner agree.[33]

Friedrich Engels, influenced by Bachofen's thinking, postu-lated in *Origin of the Family, Private Property and the State* that before they were tormented by civilization, humans lived in a state of communal bliss, punctuated by happy promiscuity and a full equality of the sexes. But the invention of animal husbandry led to the accumulation of private property by men. With male ownership of private property came a male desire for system-atic inheritance, which led men to control women's sexuality as a means of clarifying paternity. Once they possessed prop-erty, in other words, men wanted to be sure who their real heirs were, and so private property led directly to the subordination of women, "the world historical defeat of the female sex."[34]

Historian Gerda Lerner has conceptualized a progression not very different from Engels', although Lerner offers a more com-prehensive scenario and prefers to imagine a gradual institu-tionalization of male dominance over a period of some 2,500 years, between 3100 B.C. and 600 B.C. in the Middle East. But the earliest Mesopotamian documents suggest that patriarchy, or "deeply rooted patriarchal gender definitions," began before written history;[35] and so Lerner is forced to supplement her his-torical analysis with anthropological speculations about con-temporary hunting and gathering societies. In those societies, the historian declares, we find "many examples of complemen-tarity between the sexes and societies in which women have relatively high status."[36] Hunter-gatherers therefore provide the best evidence that some time during the historical shift to agri-culturalism, a "relative" egalitarianism was transformed into "highly structured societies in which both private property and the exchange of women . . . were common."[37]

The search for social ideals in the primitive harkens back to the pastoral idylls of Virgil and the idealized images presented by Melville, Gauguin, and Mead. In societies without writing, formal laws, or significant property, life is certainly simpler, and it seems likely that without the accumulation of property and wealth, there can be little concentration of power — and therefore, perhaps, fewer petty tyrants or "big men." But are contemporary preagricultural societies actually free of patriarchy and male violence?

Certainly some of them are not even remotely so. Among the North Alaskan Eskimo, according to anthropologist Ernestine Friedl, men regard women as potential wives and economic partners but also as "somewhat like a commodity that men can take, give, receive, and exchange with one another."[38] A teenage girl is potentially "fair game as a sexual object for any man who desires her. He grabs her by the belt as a sign of his intentions. If she is reluctant, he may cut off her trousers with a knife and proceed to force her into intercourse." In Aboriginal societies of the western Australian desert, described by anthropologists as egalitarian, only men can initiate a divorce.[39] And since, according to anthropologist Robert Tonkinson, men are imagined to be the "bosses" of their wives, other members of the community seldom interfere in instances of domestic violence. "Whatever a man's particular reputation, as perhaps brutal or excessively jealous, a wife cannot count on the support of the community at large."

In *The Forest People* and *Wayward Servants,* anthropologist Colin Turnbull described the Mbuti of Central Africa as an ideal society. Sex roles were not reversible among the Mbuti, but, according to Turnbull, men did not consider them very important. A Mbuti man "sees himself as a hunter, but then he could not hunt without a wife, and . . . he knows that the bulk of his diet comes from the foods gathered by the women."[40] Nevertheless, the men are still the political leaders of the society, and children come to associate their fathers with authority, their mothers with love. Actually, Mbuti men declare that a "certain amount of wife beating is considered good."[41]

Feminist anthropologist Peggy Reeves Sanday places the Mbuti among her group of supposedly rape-free cultures largely on the strength of Turnbull's writings.[42] She cites the male anthropologist as having declared "I know of no cases of rape" — even though his full sentence states, "I know of no cases of rape, though boys often talk about their intentions of forcing reluctant maidens to their will."[43] Indeed, later in the same monograph, Turnbull reports that during a certain female initiation ceremony Mbuti men in theory must have permission for sex, but when they lie down next to a girl, "if they want her they take her by surprise when petting her, and force her to their will."[44]

Anthropologist Marjorie Shostak declares the !Kung San foragers of southern Africa to be *almost* egalitarian, with no preference given to boys or girls, men and women participating fully in raising children, and the mother approximately equal in authority to the father. "All in all," Shostak writes, "!Kung women have a striking degree of autonomy over their own and their children's lives." These women become "multifaceted adults and are likely to be competent and assertive as well as nurturant and cooperative."[45]

Multifaceted. Competent. Assertive. Nurturant and cooperative. Of course, the same general descriptors might be written about women in many parts of the world, even in the heart of Western patriarchy. More particular information from Shostak's research clarifies what sexual equality means here. Among the !Kung, Shostak has written, men more commonly take the positions of political power and influence, and the man's "somewhat greater authority" is recognized by everyone, men and women. Male initiation rites are secret while female rites are public. Some male objects, such as arrows, are polluted by a menstruating woman's touch, whereas men can never pollute women's objects.

Anthropologist Richard Lee adds to the picture. He declares rape to be rare among the !Kung[46] and considers life to present relative equality for !Kung women. There is "no support" for the concept that women here are "oppressed or dominated" or "subject to sexual exploitation."[47] But men do two-thirds of the talking and more commonly become spokespersons for a group.

Men also do most of the fighting. Lee and his team observed and otherwise noted a total of thirty-four violent hand-to-hand fights without weapons between 1963 and 1969. These were actual "assaults" happening "in earnest," usually conducted within a "hysterical" crowd of onlookers while the combatants themselves squared off to "fight in dead silence, grim-faced and tight-lipped."[48] Of the thirty-four fights, about three-quarters were initiated by men and one-quarter by women. Men assaulted both other men and women in a nearly equal ratio, whereas women attacked only other women — with the single exception of a woman who attacked her husband. Because the larger national government had established a local representative to prohibit !Kung men from using their weapons and hunting tools against each other, no one was killed during these fights. But of the twenty-two murders committed during recent memory, according to Lee, all the killers were men.[49] All but two of the victims were men as well.

We are fortunate to have the compelling autobiography of a !Kung woman, Nisa, as told to Marjorie Shostak *(Nisa: The Life and Words of a !Kung Woman),* available to teach us more fully what relative egalitarianism feels like in this foraging society. It does not, unhappily, feel like freedom from male violence and threats of violence. Nisa in particular was subject to her husband's jealous rages. One time her husband beat her ferociously with a stick that he had previously cut and set aside until it dried into an especially hard implement. "He yelled, 'I'm going to beat the beauty out of you. You think you are so beautiful, that you are a beautiful woman and that I am an ugly man. Well, today I'm going to destroy all that beauty.'" He grabbed Nisa by the arm and beat her — "my back, my body, all over." He beat her until her back became swollen. "I cried and cried and couldn't stop the tears or the pain."[50] Another time, Nisa's husband cut her with a knife — he almost killed her, we are told. Perhaps, among the !Kung, wife beating is actually uncommon, but Nisa's daughter, Nai, was also subject to similar abuse. Married before the onset of puberty, Nai found herself "bothered" by her older husband for sex during her first menstruation. At last, one night, her husband became so enraged at her resistance he "tried to take her by force."[51] When Nai resisted, her husband pushed her down so powerfully that he broke her neck — a bone from her neck actually penetrated the flesh. Nai was killed, and when Nisa went to the village headman for justice, the headman called for a tribal hearing. The husband's defense was that his own wife refused to have sex with him. "The headman said [to the husband], 'You fool. When a young girl has her first menstruation, you don't have sex with her. You wait until she has finished with it. Nai knew what she was doing when she refused you. Yet, you went ahead and killed her!' Then he said, 'Tomorrow, I want you to take five goats and give them to Nisa.'"[52]

This sort of a life may seem like relative egalitarianism to some and perhaps an example of freedom from the corruption of Western patriarchy to others. To us it looks more like a familiar story that might be told almost anywhere else, pre- and post-agricultural: domestic abuse, battering, and rape, leading to murder or manslaughter, and the legal presumption that a woman's life is worth . . . five goats.

Feminists everywhere revile the extent to which humans live under and endure the distorting and unhappy control of a patriarchal civilization constructed largely by men for the ultimate purpose of controlling women or women's sexuality.[53] Traditionally, the attitude of feminist thinkers hoping to weaken the grip of patriarchy has been that this system must be a cultural invention. But a new philosophy has emerged in the last decades, an evolutionary brand of feminism that sees the emergence of patriarchy as an intimate part of human biology. Evolutionary feminists, writers like Patricia Gowaty, Sarah Hrdy, Meredith Small, and Barbara Smuts, agree with traditional feminists about the evils of patriarchy, but they do not disconnect humans from their biological past. The logic of evolutionary feminists appreciates the rich details of patriarchal history as recounted by historian Gerda Lerner, but it simultaneously rejects the notion of plumbing the human condition through reading merely the last 6,000 years of history.

Evolutionary feminists would remove our inhibitions about examining animal behavior as a technique for thinking about human behavior. They would insist that people can think about the evolutionary pressures that elicit rape, for example, or other forms of violence, without necessitating any absurd pronouncement that because rape is "natural" it is in any way forgivable. After all, no one considers the case of the black widow spider, who kills and eats her male counterpart after mating, to mean that murder and cannibalism are okay. Any behaviors can still be studied as biological phenomena, regardless of how unpleasant they are.

Despite the admirable intentions of those who believe that patriarchy is solely a cultural invention, there is too much contrary evidence. Patriarchy is worldwide and history-wide, and its origins are detectable in the social lives of chimpanzees. It serves the reproductive purposes of the men who maintain the system. Patriarchy comes from biology in the sense that it emerges from men's temperaments, out of their evolutionarily derived efforts to control women and at the same time have solidarity with fellow men in competition against outsiders. But evolutionary forces have surely shaped women, too, in minds as in bodies, in ways that both defy and contribute to the patriarchal system. If all women followed Lysistrata's injunctions and refused their husbands, they could indeed effect change. But they don't. Patriarchy has its ultimate origins in male violence, but it doesn't come from man alone, and it has its sources in the evolutionary interests of both sexes.

Still, patriarchy is not inevitable, as we will see.[54] Patriarchy emerges not as a direct mapping of genes onto behavior, but out of the particular strategies that men (and women) invent for achieving their emotional goals. And the strategies are highly flexible, as every different culture shows.

We will return to these issues of the ultimate origins and flexibility of patriarchal systems. But for the moment, our goal is simpler. We wanted to know if humans are sufficiently consistent in the tendency for male violence to provide a meaningful comparison with chimpanzees. The answer is: yes.

We cannot afford to be distracted by that old false disjunction, Galton's Error, with nature pitted against nurture. Clearly, the human condition is a consequence of both. We could find comparisons that show experience makes men violent, and we could find others that show genes make men violent. Both views are right. So let us avoid the question of whether nature or nurture is more influential, and turn to wondering why we are as we are. Though every society has its own distinctive responses to and prohibitions against it, male violence has been a human universal. Why? We have been given strong clues, ones that we can take back to the animal world we come from.

And yet, what makes the similarities between chimpanzees and humans so striking is that the pattern our two species share is certainly not true of all animals. In some species, females are more violent than males. In others, males are barely violent at all. So the human problem doesn't solve itself merely by ascribing violence to males universally for all species. What gives our own species such temperamentally violent males? Perhaps there is something odd about the line we come from. Does it matter that we are apes?

7

RELATIONSHIP
VIOLENCE

I N SEPTEMBER 1980 the world's largest captive colony of
chimpanzees consisted of four adult males and nine adult
females at the Arnhem Zoo in the Netherlands. They shared
a big island, had excellent care and lots of food. But even so, their
social life was far from relaxed. Just like chimpanzees in the
wild, the males fought to be number one. In a cycle that had
been going on for at least four years, each of the three top males
had already been the alpha at least once, and each had been
deposed after the other two ganged up against him. Individual
bravado was important, and so was the support of females. But
nothing mattered so much as the alliances among the top three
males. Whoever the alpha was, it never took long for the two
disempowered males to strike up a friendship and defeat him.
One of those two would then become alpha, and he would then
be defeated in his turn. It seemed an eternal triangle.

Chimpanzee Politics was how primatologist Frans de Waal
described this system.[1] The ape struggle for power was indeed
political. And like human politics, it led to violence when negotiation failed.[2] In July and August of 1980 the alpha male was

Luit. Luit's reign was so uneasy that the tension was visible. It was detectable in the way the males eyed each other, but mostly it could be seen in the coalitions, which had never settled into a stable pattern. The younger of Luit's two rivals, Nikkie, seemed to be trying to ally himself with Luit by being ingratiating and supportive. Nikkie would literally grovel in the dust before Luit, and joined with him in displays against females. But the third male, Yeröen, barely even acknowledged Luit's status (managing only the occasional soft grunt rather than Nikkie's ignominious self-debasement), and he added insult to injury by often trying to sit and groom with Nikkie. So Yeröen, de Waal's team of observers concluded, was seeking an alliance with Nikkie. He got it. And although their political breakthrough to unity wasn't actually seen, its effect was. The eternal triangle ended when Nikkie and Yeröen combined in a ferocious attack on Luit.

On the night of September 12, according to de Waal's description in a later publication, the males' sleeping quarters "turned red with blood."[3] In extraordinary testimony to the deep bonds among chimpanzees, by the time zookeepers discovered the evidence of mayhem early the next morning, the three males were virtually inseparable. Luit had been defeated and had already acknowledged the superior power of his combined rivals. But his subordination came too late. He had been severely wounded all over his body; despite emergency surgery, he died that evening.

On the cage floor were found several of Luit's toes and fingernails — and both testicles. Nikkie was unhurt. Yeröen had some superficial wounds. This was no accident. The evidence pointed overwhelmingly to a sustained attack in which Nikkie and Yeröen had combined in an act of — what would we call it among humans? — assassination. And Nikkie became alpha once more.

Pride, ideology, or belief restrains many people from viewing *Homo sapiens* as just another primate species, one among many. Humans have language, religion, morality — culture. Humans are able to discuss what it means to be human. Humans have big brains. God created humans to be a species separate and distinct from all other species of the natural world. Humans are unique.

Biological studies indicate a more complicated picture. We may be unique, but so is every other species, and for most of our evolution as primates, whatever was unique about the human line wasn't anything human. After all, only within the last 2 million years had our ancestors acquired brains large enough to count for inclusion in the genus *Homo*. It was only around 130,000 years ago that "full" humanity was achieved (the emergence of *Homo sapiens sapiens*, the subspecies that we call ourselves). And it was not until after 35,000 years ago that art exploded onto the archaeological record in the form of cave paintings and bone carvings.[4]

Before 2 million years ago our forebears certainly weren't human. They were still woodland apes, fascinating beings who would have been lovely in many ways, but definitely apes. Further back, between 5 and 25 million years ago, they were still apes, but in the rainforest. And still further back, in the rainforest between 25 and 65 million years ago, they were something more elusive, part of a group that gave rise to both monkeys and apes. That earlier ancestry is still only vague, its form hinted at by a few fossils and some of the primitive forms of living primates. But at least we know that as long as primates existed, our ancestors have always been primates.

Primates began in the dawn years of the Cenozoic era, 65 million years ago, when the asteroid collision that killed off the last of the dinosaurs cleared the arena for radical evolutionary experimentation. Mammals had already existed well before that moment, but now, with many of their competitors suddenly gone, they were presented with new opportunities everywhere, on land and water, in trees and on the ground. It was to the trees, to eat fruits and insects, that these earliest of our primate ancestors turned.

They weren't much like apes, of course. Our best guess is

that they were something like an opossum or a bush baby: in other words, about rat-size, active mainly at night, happy to eat fruits or gums or large insects, but nothing starchy like leaves or seeds, and largely solitary although capable of distinct sociability. These grandmothers and grandfathers of us all had grasping hands (good for holding on tight), forward-facing eyes (giving binocular vision and excellent depth perception), large brains generally more specialized for sight than smell, and (compared to other mammals) middling to high intelligence. They passed these characteristics down through countless generations to the two hundred species of apes, monkeys, and prosimians that make up today's living primates.

Among the characteristics passed down was a fairly typical mammalian set of aggressive behavior patterns. Many primate species defend their territories ferociously, and this defense is carried out by females more often than by males. Boundaries are defended first by rallying cries and then as needed by charges at the enemy and by chasing and grappling and biting. In some monkey species the fight escalates to where a group of females lines up as a tight phalanx, warriors moving shoulder to shoulder, snarling and lunging and screaming at the opposing phalanx only a few inches away. Battle lines form and re-form, isolated encounters occur at the edges of the main action, and the troops may fight for an hour or more until exhausted or until the weaker yields. Only terrestrial primates can fight with a coordination as tight as this, species like rhesus monkeys and vervets and savanna baboons. Arboreal primates find the possibilities for cooperation constrained by the drop-offs and dead ends and pathways offered by branches. But whether terrestrial or arboreal, territorial fights can be frequent and fierce, occupying up to half an hour a day for unfortunate troops that happen to find themselves with pushy next-door neighbors.

Fierce and frequent though it may be, however, this aggression is very different from the lethal raiding of chimpanzees. The goal in these fights over land or status is merely the opponent's defeat. Dominate the other group. Remove them, perhaps. But once they give up, let them go. Don't try to kill them. Most primates are satisfied with seeing the rear ends of their opponents.

The same applies to fights inside the social group, where the most frequent aggression is between rival males. In most primates, males fight more intensely than females. They accumulate more scars, for instance. But like human boxers in a ring, each male's aim is his rival's defeat, not death. Sometimes, of course, there's an accident, and so defeat coincidentally happens to be mortal. In the middle of a fight, a monkey comes crashing from a branch and breaks his neck — just as a losing boxer sometimes tragically dies. Or when two bull elephants fight and a slip on soft ground accidentally results in the sudden lethal thrust of a rival's tusk. Or when two moose fight to the point of exhaustion and the loser dies a day later from accumulated injuries.

The accidental nature of these instances emphasizes the oddness of chimpanzees and humans, with their *deliberate* searches for victims, their killing and mutilation of a helpless neighbor despite his appeals for mercy. Only for these two species is the loser's death part of the plan.

So in this important way chimpanzees and humans are exceptional when compared to the extended group of primates. However, if we ignore most of the primates and restrict our comparison just to the great apes, in some ways our patterns of violence are not so odd. It's still true that only chimpanzees and humans regularly kill adults of their own kind. Chimpanzees and humans also share other evils: political murders, beatings, and rape. It seems remarkable, therefore, to learn that rape is an ordinary act among orangutans, whereas it is unknown among most species of primates and other animals. And there's other violence to be found in the lives of apes. Male gorillas kill infants so often that the threat of violent death shapes the very core of their society. These patterns are not unique to the apes, but the intensity and range of violence make us wonder: Is there something about the apes that specially predisposes them to violence?

ORANGUTAN RAPE. That male orangutans regularly rape must be one of the best-kept secrets in the literature of popular zoology,[5] and, like much about orangutans themselves, it is still poorly understood. But it holds enormous interest, and for good reason. The occurrence of rape as an *ordinary* part of a species' behavior implies that it is an evolved adaptation to something in their biology, and this raises the frightening question of whether human rape may also be adaptive — a fearful idea because, some people worry, it whispers an excuse for evil. But even if animal parallels tell us about ourselves, they justify nothing.[6] Moreover, there are significant differences between human and orangutan rape, and one of them makes orangutans seem to come from a different world altogether. In this strange, red-haired, tree-dwelling ape species, most rapes are done by a special kind of male that increasingly appears to be a freak of the ape world: an adult male frozen in an adolescent's body.

Orangutans spend most of their time alone in the high crowns of large trees in the rainforests of Borneo and Sumatra, and when they move, they often do so slowly. Patience and a willingness to put up with constant leeches are required in equal measure of those people who would watch these apes. But even if orangutans move exasperatingly slowly, their pace has the advantage of sometimes enabling a scientific observer to keep track of them for weeks at a time. It still takes several years of following them, often looking almost straight up as much as fifty meters, before a large enough sample of encounters can be gathered to offer insights into their social relationships. The records we now possess come from a range of habitats: from pure forests to those including a mosaic of upland heath, the magnificent lowland dipterocarp, and the grueling peat swamp. They come from the flat coastal reaches of Tanjung Puting National Park in southern Borneo and the Kutai Reserve in eastern

Borneo, and from low mountain slopes in both Gunung Leuser National Park in Sumatra, and Gunung Palung National Park in western Borneo. And all the data show that the orangutan system is basically the same everywhere. Orangutans are by far the least social of the apes. Mother-infant pairs (or triads, mother with infant and juvenile offspring) are the only stable and obvious social unit. Infants are entirely dependent on their mothers for several years. Offspring stay with their mothers until early adolescence, around ten years old. And for most of the eight years between births, a mother has no sexual interest in males.

As adolescents, female orangutans demonstrate a lot of sexual curiosity and playfulness. They will masturbate, investigate the vagina with toes or objects, and try to initiate intercourse with older males even though males show little interest. But once a female has a baby, her sexual interest fades completely, probably until the time when she starts having her menstrual cycle a year or two before the birth of her next infant. We cannot, however, assume that she has sex only when she is fertile. Females of many primate species sometimes copulate when they are infertile, and even when they are pregnant. Orangutans probably do so, too.

Like humans, orangutans are choosy about their friends. Some individuals like each other, and others clearly do not; so it is not surprising that the nature of their sex-play varies a lot. In a relaxed pair, sex takes on a languorous, erotic quality. Matings can begin with oral or manual manipulation of the partner's genitalia, initiated by either male or female.[7] When they finally engage in intercourse, the couple often do it face to face: missionary style. High in the trees, a female commonly hangs from an upper branch, reclines against a lower one, in coitus for a humanlike average of eleven minutes, and sometimes up to half an hour.[8]

So in many ways their sex lives seem admirable. But there is also coercion; and to understand the nature of orangutan sexual coercion, we need to introduce the two types of males. The males that orangutan females appear to prefer, the ones we have pictured so far, are the big ones. Weighing an average of 90 kilograms, they dwarf the 40-kilogram females.[9] Big males are classic adults in the mammalian mold: loud, aggressive, and showing specific markers of their adult maleness. Adult male orangs, for example, have heads exaggerated with a high crown of fatty tissue and widened by flanges of fat jutting out from the cheeks rather like a baseball catcher's mask. Their faces are further marked by full beards and, among some individuals, partially bald heads. A large sac or pouch at the male's throat, partially inflated, serves as a resonating device for the one vocalization of orangutans audible at a significant distance, the big male's long call. Big males show other tendencies for producing loud noises, such as the habit of pushing over huge dead trees as they move through the forest, thereby registering their whereabouts with loud and dramatic crashes. And they are entirely intolerant of each other. Whenever two big males cross paths, either one chases the other, or they grapple in gladiator-style combat. These can be very violent fights, and they probably account for many of the severe injuries and scars characteristic of most big males: broken fingers and toes and teeth, missing eyes, torn lips. Big males "are distinguished as a class," we are told, "by their disfigurements."[10] Females sometimes approach these big males after hearing them, behavior that suggests that they find the males attractive.[11] They may travel with them in consortships (roughly comparable to honeymoons) that last for days, and they do appear to be willing partners. Females having sex with the big males tend to look relaxed, engaged, and willing.

Then there are the small males. They are the size of adult females and they look just like younger males at the start of puberty except for one remarkable fact — they are adults. After reaching the size of a female, these small males can stay small for a very long time, as long as eighteen years in captivity. Their testosterone levels suggest that they are perfectly capable of reproduction, and they have occasionally been seen (in captivity) to undergo a sudden growth spurt and turn into big males.[12] There is some indication that small males stay that way in response to the presence of a big male nearby. Other evidence, though, shows that even in the presence of a big male, the small males can grow big. Of course, when scientists started watching orangutans they saw the small males but didn't know how long they had been small, so the small males were mistaken, at first, for merely young males. But over time it has become clear that the small males are not necessarily young at all. Many, or even most, small males are actually full adults caught in a biological time warp. But they don't act like the big adult males. They don't give long calls. They don't make big noises. They don't show the signs of fighting. They do, however, rape females.

In June of 1968 British researcher John MacKinnon established a camp on the Segama River in northeastern Borneo and began the first successful study of orangutans. Packing only essential gear, MacKinnon slept at night on the forest floor, beneath nests where his subjects were sleeping, and followed them until he lost them or ran out of food, up to ten days at a stretch. "Heat, high humidity, rainstorms, floods and gales added to the discomfort and hazard of fieldwork," he recalled in 1974. "Leeches, wasps, mosquitos, horseflies and ticks added further problems and bears, wild pigs, snakes, crocodiles, elephants, [and] banteng . . . also produced anxious moments."[13] Over a total period of fifteen and a half months, however, MacKinnon located at least 200 orangutans and observed them for more than 1,200 hours.

During this pioneering research, MacKinnon witnessed eight matings, of which seven were, in his belief, instances of "unwilling females being raped by aggressive males." MacKinnon described them in the following way: "The females showed fear and tried to escape from the males, but were pursued, caught and sometimes struck and bitten. Sometimes the females screamed; their dependent young always did so, biting, pulling hair and hitting at the males during mating. The male usually grasped the female by her thighs or round the waist with his prehensile feet, but by pulling herself about by her arms a female can keep moving and the male is forced to follow. One mating started at the top of a tree and ended on the ground. Such rape sessions lasted about ten minutes."[14]

MacKinnon's descriptions fit earlier accounts of forced copulations in captivity, and they were soon supported by every orangutan observer who saw sex in the wild.[15] From late 1971 to late 1975, for instance, Biruté Galdikas, assisted by her husband and an additional researcher, watched a total of 58 orangutans for nearly 7,000 hours of direct observation in Indonesia's Tanjung Puting Reserve of southern Borneo; during that time they witnessed 52 matings or attempts. About a third involved an element of rape. "A female's struggles ranged in intensity and duration all the way from brief tussles with squealing and some pushing and slapping at the male's hands to protracted, violent fights in which the female struggled throughout the length of the copulation, emitted loud rape grunts, and bit the male whenever she could."[16]

At the Kutai Game Reserve in southeastern Borneo, John Mitani watched orangutans during sixteen months, from July 1981 to October 1982, and witnessed 179 matings, of which 88 percent were "forced." These forced copulations, according to Mitani, "involved protracted struggles between females and males," during which "females whimper, cry, squeal and grunt," while males would "grab, bite or slap females before they could copulate. While thrusting, males continued to restrain struggling females by grasping their arms, legs and bodies."[17] And in Ketambe, Herman Rijksen saw 58 copulations, of which 27 were judged to be rapes.[18]

Biruté Galdikas's autobiographical account of two decades among the orangutans of Borneo, *Reflections of Eden*, appeared in 1995. Although her earlier scientific writings described rape among orangutans in very forthright terms, in addressing a popular audience Galdikas chose to temper her language and refine her meaning. Thus, in *Reflections*, Galdikas relates several instances where young males force themselves sexually upon unwilling females, but these acts she now describes as rape only after extended and careful qualification. "In effect," Galdikas declares after relating one assault, the male "had committed date rape." She continues: "I do not want in any way to trivialize date rape among human beings. I know from friends and acquaintances how traumatic, and how alarmingly common, it is. But sex does not have the same meaning for orangutans that it does for humans. We view sex through cultural and moral lenses."[19]

The orangutans in Galdikas's book peer out from high, hidden places in the rainforest, observing their earthbound scientific observers as strange fellow apes: simultaneously similar and different enough to inspire both fascination and fear. Wild orangutans nearly always retained and respected an invisible barrier between themselves and humans — but a powerful young male orangutan named Gundul, born in the wild, taken captive, now living free at the research camp as part of an ex-captive release project, had been around people long enough to have lost his fear of them. One day Gundul attacked and raped an Indonesian cook at the camp. Galdikas describes this event in detail. "I had never seen Gundul threaten or assault a woman, although he frequently charged male assistants. The cook was screaming hysterically. I thought, 'He's trying to kill her.'" Galdikas, after calling for help, fought the ape with every ounce of her strength, beat him with her fists, attempted to ram a fist down his throat, but with no effect. "I began to realize that Gundul did not intend to harm the cook, but had something else in mind. The cook stopped struggling. 'It's all right,' she murmured. She lay back in my arms, with Gundul on top of her. Gundul was very calm and deliberate. He raped the cook."

Fortunately the victim was neither seriously injured nor stigmatized. Her friends remained tolerant and supportive. Her husband reasoned that since the rapist was not human, the rape should not provoke shame or rage. "Why should my wife or I be concerned? It wasn't a man."[20]

Defining rape as a copulation where the victim resists to the best of her (or his) ability, or where a likely result of such resistance would be death or bodily harm (to the victim, or to those whom she or he commonly protects), in 1989 researcher Craig Palmer surveyed the literature for cases of rape among mammals.[21] He found rape to be routine among only two species of nonhuman mammals: orangutans and elephant seals. In addition, he uncovered reports of occasional rape from studies of three other species. Jane Goodall has described rape among Gombe chimpanzees, and attempts at rape have been noted among captive gorillas and wild howler monkeys. Palmer, incidentally, had no particular bias in favor of finding or not finding rape in particular species, and he did not even draw attention to its concentration among apes. The skewed distribution of nonhuman mammalian rape — four of the five known cases for mammals occurring among primates, three of the five in apes — suggests that apes are an unusually violent species, while it also shows that a few other nonhuman mammals do rape.

Evolutionary theory suggests that any behavior occurring regularly or consistently has a logic embedded in the dynamics of natural selection for reproductive success. How could rape increase reproductive success? There is a blindingly obvious and direct possibility: By raping, the rapist may fertilize the female. Rape, in other words, may be the way for some males to achieve conception, with no other biological significance.

This looks right for some species. Outside the mammals, there are a few species in which rape or forced copulation occurs regularly. The best known is the scorpion fly, a fly with speckled wings common in the woods of southeast Michigan that feeds on dead or dying arthropods, frequently by stealing them out of spider webs.[22] Scorpion flies don't have social relationships with each other beyond an instantaneous interaction. They don't remember each other. The only obvious impact of a rape is what happens at that very moment: The female may be fertilized. Rape in scorpion flies may well be a way for low-quality males (those who can't "seduce" the female, as the more successful males do, with nuptial gifts of moribund arthropods) to have a chance at fatherhood.*

Several species of ducks, such as the common mallard, do something regularly that looks very much like rape. Males attack and try to copulate, while females fight back so vigorously they have even been drowned in the process — and ducks are not known for drowning. If rape were a fertilization tactic for these species, then one would expect the victimized females to be limited to those carrying eggs ready to be fertilized. But they are not. Males often rape females that aren't ready to lay eggs. Protagonists of the view that rape has evolved as a fertilization tactic argue that such cases are merely unimportant errors by the male, reflecting imperfect adaptation or risks worth taking. As long as the costs of making mistakes aren't very high, they say, rape is favored because the benefits of fertilization, in terms of reproductive success, are so high.[23]

Following this line of thinking, conventional wisdom would suggest that rape is a fertilization tactic in orangutans, too.

---

* "A rape attempt involves a male without nuptial offering (i.e., dead insect or salivary mass) rushing toward a passing female and lashing out his mobile abdomen at her. On the end of the abdomen is a large, muscular genital bulb with a terminal pair of genital claspers. If the male successfully grasps a leg or wing of the female with his genital claspers, he slowly attempts to re-position the female. He then secures the anterior edge of the female's right forewing in the notal organ . . . , a clamp-like structure formed from parts of the dorsum of the male's third and fourth abdominal segments. Females flee from males without nuptial gifts. If grasped by such a male's genital claspers, females fight vigorously to escape. When the female's wings are secured, the male attempts to grasp the genitalia of the female with his genital claspers. The female attempts to keep her abdominal tip away from the male's probing claspers. The male retains hold of the female's wing with the notal organ during copulation, which may last a few hours in some species." (Thornhill [1980]: 53.)

Mostly it seems that small males are the rapists, and most copulations by small males are rapes. Why? Well, small males don't give long calls, so they can't attract females from a long distance. They don't fight big males, so they can't protect females from potential aggressors. And there's little evidence that females follow small males or are otherwise attracted to them. So the small males are at a big disadvantage, it would seem, in the competition to mate. However, they can do one thing better than the big males. In the rainforest canopy world of wild orangutans, the big males are seriously inhibited by gravity. They can't keep up with the females. Simply because they are so heavy, these males appear compelled to climb slowly to avoid risking a fall. In other words, females can easily escape from a big male if they choose to. But small males, who are the same size as adult females, travel as fast as the females.

Thus, it is reasonable to theorize that small males are unattractive to females, but that their smallness allows them to avoid the big males and keep up with the females; and that natural selection has favored rape as a way for those small males to impregnate females. But that idea is still only a Just So Story — it fits, but hasn't been tested. For one thing, observers can't tell when female orangutans are fertile, so whether rape is normally directed toward fertilization is unknown. We do know that sometimes it isn't. It has been documented that in captivity a small male once raped an adolescent male, "playfully," in the language of Herman Rijksen.[24] Again, of course, such unusual instances can be dismissed as error in a system otherwise effective as a male mating tactic. Only the prejudiced can take a strong view on the question at this point, because as yet we don't know how often rapes produce babies.

This is an emotional issue, obviously, and particularly if we cross the bridge of analogy from orangutans to humans. Some assert with intense emotion that what happens among other species is irrelevant to rape among humans; others declare that precisely that emotion indicates the biological arguments are resisted from a deep fear of the truth. It is possible, however, that both sides are wrong. No one can say at present whether orangutan rape increases a male's probability of conception, or why it pays for a female to resist. And the strange, skewed distribution of rape among species has not been explained either. Given our state of ignorance, we should surely be open to alternatives.

The most plausible alternative to the fertilization tactic theory of rape is the sexual coercion hypothesis recently proposed by Barbara Smuts and Robert Smuts.[25] In some species, according to this line of thought, rape may be an evolved male mechanism whose primary aim is not fertilization in the present, but *control* — for the ultimate purpose of fertilization in the future. So, rape has entered some species' behavioral repertoire because it can increase an individual male's success in passing on genes to the next generation (as all evolved behaviors ultimately must). But the immediate purpose of the rape is not necessarily fertilization. Instead, much as many feminists have long argued, it may be domination. Applied to orangutans, this hypothesis would say that the orangutan female learns the rapist's power to control her. If successful, such domination would mean that at some time in the future the female will be more likely to fear and, out of fear, accept the male, giving him more predictable sexual control over her, especially during those times when she is most fertile. Therefore, the rape will benefit the male reproductively in the long term, even if it does not immediately.

Again, this is a Just So Story. It's merely a hypothesis suggesting interesting lines of research. We don't know whether rape makes female orangutans more likely to copulate with the rapist when they meet again. Even if this question cannot be resolved yet, however, it is worth thinking about because it broadens our vision. Much human rape occurs between individuals who know each other. The traditional underreporting of date rape and marital rape makes their significance elusive, but it's clearly possible that such rapes have a logic in terms of the dynamic of the social relationship. Rape, by reminding a man's partner of his physical power, may increase his sexual control over her. Society's growing sympathy for rape victims may thus be working to end a system that has deep evolutionary roots.

To take this line of thinking further, relationship dynamics may help explain even rape by strangers. Consider the kidnapping of women by men from a neighboring tribe or village or community, or the type of war rape where soldiers move into a village and stay for weeks. The man demonstrates that he can have sex with his victim whether she wants it or not, so it might be in her own interests to accept him — much as in male prisons a victim sometimes comes to accept his rapist as a partner and protector. By a logic that challenges our strongest moral principles it could pay the woman to acknowledge the rapist's power and form a relationship that, while initially repellent, she comes to accept. This same sort of unpleasant bind we will shortly see played out among gorillas, where males make themselves attractive by killing a baby. In both cases, a demonstration of power implies that the female's safest future is to bond with the violent male.[26]

Among orangutans, rape accounts for one-third to one-half or more of all copulations. Even among chimpanzees, where rape is a good deal rarer, it probably still happens as often as among many human populations. The life histories of gorillas — and bonobos as we will later describe — show that rape is not inevitable if you are an ape. Nonetheless, rape occurs much more commonly among the great apes than among most animals. Why is rape so common for this group?

Part of the answer comes from examining social systems. Even though rape is unusually common for apes generally, each of the five species shows a distinctive distribution for the behavior. Males can't rape if society prevents them. Rapists can be stopped by social alliances. For humans as for orangutans, a sufficient mob of supporting kin or others could stop a rapist. And it is precisely a mob that the orangutan female lacks. Female orangutans live alone. Female gorillas, on the other hand, live in troops that are protected from strange males by the females' own strong bond with a chosen mate — the silverback —

who, because there are no rival males to steal copulations, leaves the timing of mating to the female. Gorillas living in troops are safe from rape. Orangutans, because they live alone, are vulnerable.

CHIMPANZEE BATTERING. The patterns of rape in the five ape species suggest this strong idea: Safety is found in numbers. This could be a clue with wider significance. If vulnerability breeds sexual coercion, grouping and alliance may help explain other patterns of aggression.

At Gombe, when a male chimpanzee reaches adolescence he initiates a startling sequence of behaviors — a ritual, virtually — that has no equivalent among females. As a juvenile he would occasionally tease an adult female with mock aggression, and sometimes he would be chased for his pains. But now, as the young male reaches the size of an adult female, he suddenly begins to do more than tease. He enters the world of adult males by being systematically brutal toward each female in turn (when adult males aren't close enough to take sides) until he has dominated all of them. As one would expect, females appear to resent this change immensely. Initially, they refuse to acknowledge the young male who, instead of behaving like the juvenile they knew, has suddenly adopted the airs of an adult. But the young male has gotten bigger, and it doesn't take more than a few encounters for the females to accept the change. The male's challenge is expressed unambiguously. In a typical interaction, he might charge at the female, hit her, kick her, pull her off balance, jump on top of her huddled and screaming form, slap her, lift her and slam her to the ground, and charge off again. This never becomes the all-out sort of wild gang violence directed against males and sometimes females from other communities during lethal raiding. The female is not killed, rarely injured. But it is still plenty brutal. Pretty soon she meekly approaches him — and he might ostentatiously extend a hand, from a shoulder held high, to touch her with proud reassurance.

Nor does it stop there. In subsequent years, the male will

often attack females without apparent provocation and with similar ferocity.

Sexual coercion looks to be the underlying reason, because mating patterns in Gombe demonstrate how particularly effective male domination is. A sexually receptive female will attract a flock of males who follow her with intense attention. These females clearly favor some males more than others — but their choice is limited. As Goodall describes it: "Almost always, unless he is crippled or very old, an adult male can coerce an unwilling female into copulating with him."[27] So the rule seems to be: Coercion works.

But a single copulation is probably not the male's most important reproductive benefit from coercion. By exerting his control, the male can also sometimes force a female into a consortship. A chimpanzee female is most fertile during the last two to three days of her monthly menstrual cycle, a time marked by dramatic ano-genital swelling. And for those two to three days aggressive competition among the males often becomes overwhelming, especially if the community's alpha male is not supremely powerful. Males routinely challenge each other, and in the course of these challenges the female often suffers, whether from being herded or from running to avoid clashes and chases among the males. Fertile females find little time to eat, so they sometimes end up being forced to delay satisfying their hunger until nighttime, feeding for an hour or more after the males have built their nests and turned in. They regularly receive wounds, many probably from accidents occurring when they are chased. Or they can fall out of trees and drop fifteen meters or more. And their offspring, still needing their mother at three to seven years old, seem alarmed or lonely, but are kept away from her for most of the day by the danger of being caught up in the middle of a fight. It is certainly a highly stressful time.

Perhaps that stress is one reason why a female will often disappear from the center of the community for several days, or even a few weeks, when she is cycling. She leaves with just her offspring and a single male and they travel together quietly on the edge of the community's range. Of course, this makes hard work for human observers, who may have a two-hour walk across difficult terrain to reach a consorting couple before they leave their nests at dawn. But it's worth it to see the contrast, the relaxed, easy life of a family-like party traveling just a short distance each day. And current data suggest that at Gombe infants are especially likely to be conceived during these honeymoons or consortships.[28]

In evolutionary terms, consortships are exceedingly valuable for the male not just because the female is likely to conceive, but also because he has no rivals for fatherhood.[29] So one cannot be surprised to learn that males use many clever devices to induce them. Some of the devices are friendly. Males who groom females often and share meat with them are regular consorters; but if friendliness doesn't work, males frequently turn to violence. He may start out by trying to catch her attention in a big group. Perhaps he grooms her for a long time, continuing to do so even when the others leave for the next fruit tree. Then, after the others have all gone one way, he walks a few paces in a different direction and turns to look at her. A significant look. She doesn't follow. He tries again. She starts in the direction taken not by the male but by the rest of their party. So he attacks her. She screams, but the other males may be out of hearing distance, or, if it is early in her cycle, they may not care. He sits and waits for her to follow. Again she walks away from him. Again he attacks. After several such attacks, each time increasing in intensity, she follows. The result? A consortship. Signs of her having been coerced normally soon fade, and the pair may move together peacefully. But he may yet turn and attack her brutally, and sometimes she shows her unhappiness about the situation by escaping when she hears other males calling.

Male attacks on females, so consistent and regular an aspect of chimpanzee life, might best be described by the term *battering* — as it is used to describe domestic violence among humans, most often when a man attacks and beats up a woman with whom he has or has had an ongoing relationship. Chimpanzee battering and human battering are similar in three respects. First, they are both cases of predominantly male against female violence. Second, they are both instances of relationship violence; male chimpanzees batter females who are members of their community, ordinarily known to them for many years, often in contexts with nothing material, such as food or support for an ally, at stake. Third, like human battering, the battering of a female chimpanzee may take place during or be triggered by a number of superficial contexts, but the underlying issue looks to be domination or control.

Like lethal raiding and like rape, male against female battering is a rare behavior among animals generally. What explains its distribution? You might think it merely a result of males being larger than females. It's true that a male's size does help him batter, but that's only a part of the story. Gorillas don't batter, for example; although males are twice the size of females and invariably come to dominate females totally, they do so without being brutal. More important than the size difference between male and female, so a survey by Barbara and Robert Smuts suggests, is vulnerability — the same factor that accounted for rape.[30] Battering in animals occurs in species where females have few allies, or where males have bonds with each other.

GORILLA INFANTICIDE. The earliest reports on gorilla behavior were written by gun-toting men who, having placed themselves in danger by provoking a gorilla charge, shot their guns and felt some pride and justification for killing the killer they had just created. "Then the underbrush swayed rapidly just ahead, and presently before us stood an immense male gorilla," wrote the American adventurer-reporter Paul du Chaillu in 1861. "He had gone through the jungle on his all-fours; but when he saw our party he erected himself and looked us boldly in the face. He stood about a dozen yards from us, and was a sight I think I shall never forget. Nearly six feet high (he proved four inches shorter), with immense body, huge chest, and great muscular arms, with fiercely glaring large deep gray eyes, and a hellish expression of face, which seemed to me like some nightmare vision: thus stood before us the king of the African forest." The gorilla clapped his chest and roared, perhaps about to charge. Du Chaillu, gun raised, finger on the trigger, thought thus: "And now truly he reminded me of nothing but some hellish dream creature — a being of that hideous order, half-man half-beast, which we find pictured by old artists in some representations of infernal regions."[31] Du Chaillu blasted the animal, returned to America with his skin and skeleton, wrote a book, and the first myth began: Gorilla as King Kong, the ferocious beast.

A hundred years later other observers came into the territory of gorillas without guns, observed the apes patiently, and discovered a deeper reality about their lives. Gorillas are for the most part quiet, relaxed, and affectionate with each other. They resemble humans in many of their gestures and actions. They live in stable family troops typically consisting of one fully adult male — the silverback (so named because of the broad silver stripe on his back) — a harem of perhaps three or four adult females, and their offspring. Gorillas are family animals with stable mating patterns; they are devoted mothers and tolerant fathers. And the predominant impression they give is of a magnificent peacefulness. Females are totally subordinate to silverbacks. Among themselves the females rarely even bicker, and when they do, their quarrel is quickly suppressed by the silverback, who is twice their size. Sometimes a family troop will include two or more silverbacks, up to an extraordinary seven in one troop in Rwanda. Even then, however, the males compete little with each other: accepting the status quo, seldom and barely making alliances with one another to jockey for power. The gorilla harem society means that some adult males mate with many females, while many other males mate with none. A son who stays in his father's troop can sometimes inherit it. But most are not so lucky and so must make their fortunes alone without help from kin. These extra males either wander alone or join together in bachelor troops, and even in those troops, aggression is rare. Gorilla society appeared to represent yet another version of the ape Eden. King Kong was replaced by a second myth: gentle giant.

Gorillas *are* gentle and peaceful most of the time. But less than a year before Jane Goodall's team discovered the body of Godi, Dian Fossey discovered the body of a dead gorilla baby, Curry. "In April 1973," she wrote in *Gorillas in the Mist*, "when Curry was nearly ten months old, a tracker found the baby's broken body left on a flee trail after an interaction between Group 5 and a silverback. Examination of the corpse revealed ten bite wounds of varying severity. One bite had fractured the infant's femur and a second had ruptured the gut. . . . Curry was my first introduction to infanticide among the Visoke gorillas."[32]

Curry's death quickly proved to be part of a pattern. By 1989, four years after Fossey had herself fallen victim to murder in the Virunga Volcanoes, there were data on fifty infants at her research site. In the sample, 38 percent of infants died before they were three years old, and at least 37 percent of these deaths were judged to be from infanticide, or about one infant in seven, overall. The figures suggest that the average gorilla female experiences infanticide at least once in her lifetime.[33]

The danger is paramount after a breeding silverback dies, leaving a mother and her infant unprotected. In the Virungas, fourteen observed encounters between unfamiliar males and vulnerable infants led to at least six deaths. Indeed, the only case of a vulnerable infant *known* to have escaped infanticide was one who died of pneumonia two weeks after his father's death, before his mother encountered a new male. It looks as though most infants unprotected by a silverback are killed.

The mere fact that a murdered infant was fathered by another male means that the killer's genetic interests are served by infanticide because he removes the competitor's genes. But the male gorilla benefits in a second, more direct way. Females

whose infants are killed may voluntarily join the killer's troop and have their next baby with him.

This seems odd. A gorilla mother is intensely affectionate, clearly very strongly bonded with her infant. So why should she join her baby's killer? It's not as if any direct threat forces her to do so; males don't bully females directly or try to kidnap them. A female can always leave of her own accord. So she doesn't have to join the killer. She could join any of the half a dozen troops that share her neighborhood. But the very act of infanticide makes the killer attractive. In fact, so strong is this strange, counterintuitive logic that it appears responsible for a second, rarer form of infanticide. It drives males to attempt infanticide when the chance of success is very small: when the mother has a protector, the infant's father.

Consider the strategy of a bachelor male who is ready to breed. His reproductive career depends on enticing females to join him. The best gift he can offer a female is protection from infanticidal males. How does he persuade her that he'll protect her future infants well? He can show his dominance over her current mate in the most unambiguous manner. He can kill their baby.

The bachelor silverbacks are tenacious and ingenious.[34] Tiger, for example, was a bachelor who approached and followed breeding troops for months. Once he pursued an established troop, Nunki's Group, for a full week. For his pains he was chased nearly twice an hour, and he and Nunki had three fights. Tiger chose his ground carefully, it seemed. Once, when the troop was foraging contentedly on the lower part of a slope, Tiger circled above them to start his run with the advantage of gravity, so that by the time he hurtled through the troop, he was plowing down at full speed and unanticipated, like a fighter pilot diving out of the sun. Sometimes Tiger slept so close that he knew the geography of their sleeping area, and could charge in a surprise raid before dawn. Three times he chased them throughout the night. Tiger eventually acquired one of Nunki's six females, though he lost her to another silverback after four months. Later he won another female whom he kept for a year and a half before his early death.

When a male kills her infant, the female is an established member of an existing troop, while the killer male is a stranger. If she has seen him before, it has been only during violent interactions when he challenged her mate — a patent threat to her infant, a blur of power as he rammed through the vegetation before being stopped, outfought, and repelled by the resident silverback. And now he has succeeded in his aim. He has managed to get past the defenses of her mate. He has charged directly up to her, even as she screamed and fought back, and in a terrifying show of mastery, he has torn her baby from her and killed it in an instant.

Yes, for the most part gorillas are gentle giants, but the gentleness is interspersed with violence, and their apparent peace is overlain with fear. Females are trapped in a vortex of male-initiated violence. The silverback they live with is good only so long as he is strong enough to fight off all comers. When another male does break through the defenses and kills her infant, she responds in a way that violates all our assumptions about attachment, loss, and revenge. It may take a few days before the female leaves her troop, but the evidence is clear. Infanticide draws a female to a male. She leaves her old mate and joins the killer. She may mate with him, have babies with him, and spend the rest of her life with him. The female's choice is imposed by the logic of violence, by the threat to her next infant. The new silverback has become her hired gun in an ape universe of silverback baby killers.

We began with the aim of finding out whether our nearest relatives are in general given to violence. To some extent, the three we have considered are. Orangutans, chimpanzees, and gorillas are all demonic male species. In spite of the wonderful, peaceful continuity of their daily lives, adult males of these species are potential brutes, liable to those occasional acts of violence of a type and scale that in human society would count as heinous crimes. And yet, as we have seen, their violence is by no means random or general. Rather, it follows a few definite patterns within the social lives of each species. Most female orangutans are raped regularly — but there are no indications of orangutan infanticide. Every female chimpanzee gets battered, some are raped, and a few have their infants killed. Many or even most gorilla mothers experience infanticide — but they aren't battered. Rape and battering and infanticide are more than simply holdovers from deep in the past or expressions of some ancestral tendency for generalized violence. Instead, these three styles of male violence make sense within particular contexts.

The lives of orangutans, chimpanzees, and gorillas support the common-sense view that the social system of a particular species helps determine whether aggression pays off. For orangutans, while we may not yet understand why small males benefit reproductively from raping, we can at least surmise that females are likely to be raped because, commonly moving through the forest alone, they are vulnerable. For chimpanzees, battering helps a male get his way when it most matters to him; and females are vulnerable during those times when they are alone or without allies. For gorillas, infanticide is a demonstration by males of female vulnerability.

But female vulnerability is only part of the picture. Another part is the intelligence of the species. The underlying premise of relationship violence is that it works best (that is, it increases the reproductive success of the individual using it) where animals are intelligent enough to learn each other's personalities. All the great apes, we know, have especially sophisticated brains, capable of remembering human signing gestures learned while young but not used again for seventeen years, of understanding numerical concepts, of creating a simple form of art.[35] Such intellectual skills are the kind of abilities that, both in the wild and in captivity, give these species especially rich and multidimensional social relationships. They therefore amplify the range of tactics that individuals can use to interact with and manipulate each other. Some of those tactics are affectionate, and some are violent. The hugs and kisses and embraces of the apes are as elaborate as their use of brute force. In both cases, good and bad, the fine memories of these apes provide long-term meaning to their behavior. Intelligence turns affection into love and aggression into punishment and control. Far from being the mindless expression of some deep and bizarre ancestral trait then, the intense violence of apes arises partly from the very elaboration of their cognitive abilities.

8

THE·PRICE OF FREEDOM

USK ON THE PLAINS and the moon is up. The African night has changed little in the last 5 million years. Hoot of an owl. Warm breeze. Dusty scents. Fluttering moths. Every direction holds whispering sounds, uneasy silences, occasional grunts and whistles and chirps. The prey is out there. And the predators. For some, there's another possibility as well. The night could bring a fight not just with predators, but with their own.

A hunting party is on the move, four adults. The young ones stayed in camp with an aunt. The hunters have walked ten kilometers already. No easy prey yet tonight. The big herds have gone north. Maybe the only reward will be the stale meat of an old carcass. But the night is young.

Glances and pauses reveal tensions as the travelers reach a crest of gently rising ground. They haven't heard calls from others in their clan for an hour or more. They're already past the edge of familiar territory, the preferred hunting area. But Sally, the leader, wants to go onward. The others are more cautious. Susan, Sarah, and Alan stay back. Out there, the danger level rises. *Lions are likely out here. The lions will be bold, too, be-*

cause this is part of C-clan's usual hunting territory. C-clan hasn't intimidated the resident lion pride in the same way our gang has. Worse than lions, the C-clan might be there, out for prey in a hunting party like us. Relations haven't been good for the past few months. There's bound to be a fight if we meet them. No need to court danger unnecessarily. Who needs it?

Sally's party assembles, facing ahead. It's still light enough to see the shape. Someone out there, walking quietly, a hundred meters away. *Surely, it must be one of the neighbors. And she's alone!*

Sally, Susan, and Sarah pace quickly forward, alert and focused. The loner stops at the wreck of a half-eaten wildebeest, turns, and suddenly sees the invaders. Sally leads her pack in, caution aside. Closing, she recognizes the stranger. *It's Carla, that big mother with the cut ear and long laugh. She killed Susan's younger sister four years ago, ripped her apart; dead in the middle of her first pregnancy. What a pleasure it would be to get her.*

Carla sees them, turns, skips off, and is soon racing, but by the time she's reached full speed, Sally's gang is almost on her. Carla, headed off by a rocky outcrop, is forced to turn — but Susan and her adolescent daughter Sarah, lean and fit, anticipate Carla's move. Within seconds, Carla feels the first hit on her back. She bends away and squeals and races on, but it's too late. She stops to face Sally now, trying to make her back down. Suddenly she's surrounded by intense, eager, aroused bodies. Young and old, the killers' excitement is the same. Chance for a big score.

Quivering, appalled, aware, surrounded by her neighbors' blood lust, Carla can't face every way at once. Her best is not enough. Wherever she turns, there's someone behind her, able to get at her back. It doesn't take long. Sally and Sarah are the first to push in — but everyone is eager and showing off, even Alan. Within seconds Carla is down. Her screams stop quickly.

All right! A job well done. And there's food as well.

The killers are just starting on the wildebeest carcass when their elation freezes. Calls from the east announce owners of the territory: a hunting party from the C-clan. *A big party. That's annoying. A massed fight would be risky. Even without Carla on their side, the neighbors are powerful, particularly Chris and Charlene — those two always work together and fight hard. There'll be fights with the C-clan soon enough, if a kill comes down in the boundary area. Till then, what's the point?* Sally, Susan, and Sarah decide quickly. They tear off a mouthful each, then the sisterhood heads away, fast and united. Alan joins the race for his own safety. . . .

Where their prey is sufficiently abundant all year, spotted hyenas live in female-dominated clans that rule access to their hunting grounds and protect the all-important denning area. The clans fight with their neighbors both in defense of these hunting grounds and in occasional deliberate chases into the next territory.[1] Much like chimpanzees or humans, spotted hyenas live in a society of temporary associations within a stable larger community. Their daily life has much to commend it: relaxed play, idle siestas, gentle grooming, affectionate greetings. But it's also a life where power matters. Fights between clans go the way of the strongest, and they are conducted with all the viciousness of battles between rival street gangs.[2] The essential dynamic is the same. It's a game of *us against them.* Fear of the stranger unites female spotted hyenas, and affection cements their bonds with one another. And when the odds are overwhelming, the aggressors don't stop at victory.[3] They kill.

Is it natural for animals to try to kill their own species? It's certainly unusual for adults to kill adults. As we noted earlier, fighting adults of almost all species normally stop at winning: They don't go on to kill. That's what makes the lethal raids of chimpanzees and humans an interesting puzzle. But a wider look at nature shows that killing per se is not unusual. In many species, the killing of infants is routine; it's the killing of adults that's rare. A few species provide exceptions, like lions and wolves and spotted hyenas. Like chimpanzees, in certain circumstances they punish rival adults with death. Yet there are important differences among these species. In chimpanzees and humans the killing is dominated by males and characterized by raiding; for lions it's dominated by males participating in take-overs, not raids; whereas spotted hyena killings don't involve raids and are initiated by female-run gangs. The picture has suddenly become more complex. Nevertheless we can find enough in common to help answer a question that lies at the heart of the ape problem. What makes some species deliberate killers of their own kind?

Beginning in the 1960s, our understanding of animals was revolutionized by an enormous increase in the number of people who watched, rather than guessed, how animals live in the wild. Apes are merely one tiny subset of animals whose behavior has now become familiar to us. So many of the world's other larger species have been studied in the last few decades that it is now hard for an ambitious student to find one of the so-called charismatic megavertebrates whose society has not already been charted for a year or more. As the years of observation have piled up, so have the number of rare incidents, and the result is a sea change in our appreciation of the role of violence in nature. We know of animals manipulating, deceiving, attacking, and challenging each other in ways that were barely hinted at before the revolution in observing. Even rare events that may take place less than once in an animal's lifetime have become part of our scientific lore. And from all those years of watching, we know that for many species, infanticide is an extremely important part of life.[4]

When infanticide among monkeys was first witnessed in the 1960s, conventional wisdom took it to be abnormal — an accident, perhaps, or a meaningless result of human interference. These were innocent times when it came to nature, times, after all, when so distinguished a scientist as the Nobel prize–winning ethologist Konrad Lorenz could claim that well-adapted animals don't kill each other. For anthropologists in particular, the concept that anything as nasty as infanticide could be part of the ordinary workings of nature was alarming; animal infanticide violated the concept of groups evolving for the social good. But the new wave of evolutionary theory that began in the 1960s readily explained infanticide by male monkeys as a behavior that benefited the killer by increasing his personal reproductive success, regardless of its effects on others.

At first, those who argued that infanticide made sense in animal societies were vilified as reactionaries with a crude political agenda. The ivory towers of anthropology rang with the sound of fighting between the old and the new guard.[5] With the dust now settled, we know that infanticide is a behavior typical of certain species within every major group of animals. It happens in birds and fish and insects. For mammals, infanticide is rampant among rodents and carnivores and primates. And indeed, the fossil bones of a baby dinosaur in an adult's rib cage even suggest dinosaurs did it 220 million years ago.[6] Depending on the species, the typical killers can be female or male, adult or immature. And the evolutionary benefits are various: from using the infant as food to accelerating the mother's sexual availability. Far from being abnormal among animals, infanticide is routine.

Where they live in the open plains, lions can be watched for days at a time. Since they quickly learn to tolerate vehicles close to them, they are wonderful subjects for studying rare events. In 1988 a British film team in Serengeti National Park, Tanzania, followed two male lions for weeks, hoping to capture infanticide on film.[7] The team knew what to expect. Male lions enter a pride by defeating the prior male residents. Then they search for infants and kill them. After that moment, the mothers are trapped by the males' action. Any real delay in becoming fertile, now that a lioness has lost her suckling infants, reduces her own genetic output. So, because natural selection rather than a human-derived ethos has shaped her emotions, she challenges our moral senses by accepting the loss of her infants and becoming immediately coquettish with the killer males. Within days she seduces the new males, using the same playful flicks of her tail and rubbing of her curled body that a domestic tabby might show in happier circumstances. And within months she has new offspring. She has, of course, unwittingly contributed to the evolved cycle of male violence because her own reproductive strategy, in making the best of a bad job, rewards the males genetically for killing her infants. In the Serengeti, a quarter of all infants are sacrificed on the altar of infanticidal male selfishness.[8]

So when a pair of males hunted down and then chased out a pride sire, the filmmakers followed the new males relentlessly. After weeks came the painful moment, which, in spite of its horror, made Queen of Beasts one of the most remarkable films about any mammal. The cubs, when the males spy them, are a picture of innocent contentment, piled on top of each other in the short dry grass. A male first stalks with the look of a hunting dog, head up, eyes focused on the target, and body rigid except for a tense slow walk. After fifteen seconds the walk becomes a trot. The final approach is swift, unhesitating. The first cub dies within two or three seconds, the male shaking it in his jaws and then releasing the dead cub and looking around with his head strangely high, mouth open, savoring for a moment the taste. The survivors scatter, but the male takes only nine seconds to find his next victim. This time the cub knows what's coming. Watching the advancing killer, the cuddly cub, in a desperate attempt to save its life, rears, screams, and falls backward even while the male's head looms above. The cub lies for a second like a baby on its back, arms and legs spread wide in tragic resemblance of an invitation to embrace, showing its unfamiliar soft white belly. Then the cub acts as if it knows there's no chance of appeasement. It rolls onto its side, and curls an arm up as if to hide the sight. The soundtrack includes the narrator's cool, detached voice laying out the evolutionary logic as the jaws of the new male close on the infant's ribs. In a few seconds the corpse hangs from the killer male's mouth.

In species where infanticide or even the threat of it occurs frequently, the reverberations within the rest of the social system can be huge.[9] Gorillas are a perfect example, for among gorillas the very existence of their harem troops can be attributed to the females' need for protection from killer males. All through the primate order we find similar examples: species whose societies or mating patterns or friendships appear strongly shaped by the need to protect against this one terrible threat that never goes away. Protection comes in various forms. It can come, as in gorillas, by mothers bonding with a strong male. It can come, as with chimpanzees, by a female mating with as many males as possible, the result being that all the males remember that they *might* be the father and so are powerfully restrained from harming the infant. (An odd puzzle for chimpanzee specialists is that this system can go wrong, shown by a famous case in which a male killed an infant he might have fathered.)[10] It can come, as it does among langurs, from females banding together to support their male in fights against potential usurpers of the breeding role. It can come, as in lions, from females hiding their infants. It can come, as it does for many species, from mothers of new babies staying especially close to their mates. The threat of infanticide from individuals (usually males, but occasionally females) who see an infant merely as a barrier to their own goals is an abiding social pressure that many species have surely lived with, and reacted to, for millions of years. The severity and style of the threat vary among species according to the benefits that killers gain as well as the counterstrategies that mothers, infants, and sometimes fathers can bring to bear. But it is pervasively, persistently present.

We bring infanticide to the question of lethal raiding not to discuss its effects on society, but to settle a simpler question. Do animals regularly kill each other? The answer is abundantly clear: Yes, in many species they do. Animals have been selected over evolutionary time to do whatever it takes to advance an individual's own genetic interests, even when those interests look unpleasant or contrary to the well-being of the species as a whole. If killing can increase an individual's reproductive success, then it is favored. Because of its wide distribution and easy

explanations, infanticide shows more clearly than any other behavioral pattern the ultimate heartlessness of nature. Countless
animals kill infants with all the emotion of a farmer picking a
turnip.

But why, then, do animals kill infants so much more commonly than they kill adults? The answer mocks with its banal-
ity: Infants are easier to kill. Adults present a threat to the potential killer.

As we have seen, spotted hyenas, like lions and wolves,[11] provide
a rare exception to the general rule that animals don't deliberately kill other adults of their own species.[12] These predators all
forge bonds with other adults, but regardless of their bonds, they
sometimes move alone. Occasionally, their wanderings take
them into neighboring territory. When individuals are caught
alone by a hostile party, they can be chased, cornered, attacked,
and killed.

This pattern bears a strong resemblance to lethal raiding in
chimpanzees and humans. So far, however, we find no evidence
that hyenas or wolves or lions enter a neighboring party deliberately in search of enemies to kill. Also, for hyenas and wolves
and lions, females play a big part in the aggression. But those
differences aside, we're dealing here with a system that reminds
us very much of our own crimes, that speaks straight to the gut
of a human victim surrounded by confident foes. Here is the fear
you feel on an empty street when the muggers step out from the
shadows, the same quick arithmetic that may enter the minds of
criminal and victim simultaneously, the same acknowledgment
of the importance of numbers.

In their extraordinary film *Eternal Enemies*, Beverly and
Dereck Joubert follow lions in northern Botswana's Chobe National Park.[13] The lions live in prides that defend their territories against neighbors. The prides are parties of female kin, well
known to the Jouberts from a decade of observation. During a
zebra hunt one night, two lion prides converge near the boundary and a fight ensues. In the chaos, an old female of the Maome

pride becomes isolated by intruders behind the battle lines. We see her surrounded and held captive at first by three hostile lionesses, eventually by as many as seven. And her death is particularly distressing because it looks so deliberate. She starts erect, alert, snarling, though already bleeding from the shoulder. Naturally she can face only one way at a time. She turns repeatedly to check behind her. She is wise to check, for lions can die from a single bite to the spine, but whenever her head is turned, someone swipes at her. The attempt to stop one antagonist merely opens the door for another. Like some hideous children's game, everyone takes a turn striking or biting, while the surrounded victim hopelessly spins and writhes and rears and twists. She is prevented from escaping, constantly herded back to the center. Motsumi, the leader of the Maome pride, approaches once in an apparent effort to rescue her pride-mate, but she is quickly chased off. And the victim is tormented in this fashion for several hours before she weakens and finally collapses, exhausted, finished. The female killers leave, and then the corpse is eaten by hyenas. With minimal risk to themselves, the pride has relentlessly caused a rival's death.[14]

This killing calls to mind an exquisitely cruel execution devised by Aztec warriors. On special occasions these Central Americans roped a distinguished enemy warrior to a high platform, binding his legs tightly but allowing his arms considerable freedom. They gave him four throwing clubs and a sword edged with feathers. Then four champions rose to the edges of the platform, surrounded him, and, using swords edged with obsidian, flayed the enemy warrior as precisely as possible, making delicate stripey cuts all over his body until, after several hours, he collapsed.[15] The captor of this distinguished victim would later drink what remained of his blood. For the Aztecs this was a sacred ritual. For the lions, of course, surrounding and tearing slowly away at their victim is merely practical. But in either case, the kill is efficient from one very important point of view: It's safe for the killers.

For most species the problem with killing adults is the danger of being killed oneself.[16] The quick bite or whipping arm that so decisively dispatches an infant may merely wound and enrage an equal. If the defeat of a rival peer serves just as well as his death, there is no sense in risking your own life by attempting that final push. Balanced power, in short, reduces the likelihood of killing.

Numbers, like weapons, change the calculus of violence by creating unbalanced power. Yeröen and Nikkie, working as a pair during the night at Arnhem Zoo, were barely hurt in their lethal attack on Luit. None of the Kasekela chimpanzees, raiding in gangs at Gombe, was hurt in their attacks on the lone individuals of the Kahama group. The hyenas, lions, and wolves that kill their neighbors do so when loners can be incapacitated, so that the aggressors are unhurt. Of course, the same principle guides a wise military leader in our own world. *Bring massive force to bear. Otherwise, avoid engagement. Isolate and surround, or merely stand and stare.* Here is the message taught at Sandhurst, at West Point, and no doubt in small village councils throughout the land of the Yanomamö. It is also a message beautifully told by a species with more to gain from fighting than almost any other: the honeypot ant.

The honeypot ant, living in large underground colonies in the mesquite-acacia bushland of Arizona, is one of nature's stranger inventions.[17] In many ways they are just ordinary little bugs. They eat termites and nectar, and under most conditions they look and act quite like other ants. But their intercolony relationships are marked by a humanlike style of war.

Their popular name derives from a special caste of individuals within the species, the honeypots, that become a living foodstore for the rest of the colony when there is not much food elsewhere. Members of this caste, known more formally as *repletes,* cling to the ceiling of their nest chamber and are fed by workers until their abdomens swell up into cherry-size spheres. As many as three hundred of these living honeypots hang up there, patient reservoirs of liquid sweetness, ready to disgorge food in hard times when solicited by their nestmates. As a caste deliberately fed to the point of immobility, these ants may find their nearest human equivalents in the force-fed, milk-fattened young wives of the East African king of Karagwe. When they were measured by a fascinated John Hanning Speke in 1861, the wives were so incapable of movement under their own steam that they had to be rolled on the ground for him to complete his measurements.[18] But no humans can distend their bodies like the honeypots. If people clung to a ceiling and had their lower halves swell in the same proportion, their bodies would become roundish vats six feet across and weighing about eight tons. So one can easily understand that the individuals of the replete caste, once they are turned into community reservoirs, will never move again under their own volition. They can, however, be moved by other ants. In other words, the replete honeypot ant has become within its own world a stored and stealable resource of great value. Native Americans of the region relished these fat ants enough to steal them from colonies; and so do their own ant neighbors.

Neighboring honeypot ant colonies frequently forage in overlapping areas, a habit that often leads to massive territorial confrontations lasting for days and sometimes involving several hundred warriors on either side. When one colony defeats another, the victors drive forward and sack the losers' home. Rival queens are killed or chased away. Larvae, pupae, and young workers are hauled off to the victorious colony, where they become nonreproducing slaves. And finally, the prized honeypots are laboriously pried off the ceiling of the vanquished colony's nest, and then slowly shifted over to the winners' nest by workers who may take several days to drag their huge prizes a few feet.

Slaves, an enormous food supply, and the extinction of rival neighbors might seem enough spoils that the colonies would take heroic risks to attain victory in battle. Instead, their fights are some of the most timid known among ants, consisting for the most part of showing off. Scouts from rival colonies stand tall opposite each other, bend their abdomens toward each other, but usually they don't engage. Why not? Bert Hölldobler, the biologist who discovered how honeypot ants fight, notes that direct physical fights almost always end in the death of both opponents. This typical outcome occurs because honeypot ants have an unusual combination of good weapons and poor defenses. Their good weapons are standard for predatory ants: ferocious jaws, used mostly in this case for killing termites. But the poor defense is a rarity. The material of their tough outer shell is extremely thin compared to other species of ants because for some individuals it will have to stretch into honeypots. So, as with the case of mammals that pair off, fighting one on one, any direct attempt to kill doesn't pay. It's just too dangerous for both parties.

A colony will engage in battle only when it discovers a massive imbalance of power. In the ordinary sequence of their lives, some ants from the colony move around their desert territory as scouts. Upon encountering neighbors, the scouts return to their home nests in order to recruit reinforcements. The reinforcements gather and move toward the neighboring colony, which has by then organized its own fighting force. If the two colonies have roughly equal numbers and their armies are about the same size, there is a standoff. No one fights. But if one colony has a decisive numerical superiority it will attack and overwhelm its neighbors. The moment of physical combat is brief. The smaller group gives up quickly, with most workers likely to accept slavery over death.

Here is a species with larger resource prizes for victory in combat, possibly, than in any other except humans. Yet these ants don't sacrifice themselves in suicidal thrusts. They kill their neighbors, true, but they kill them when they can do so with little risk to themselves. This species, despite its human-like raids for resources, its robotic control of fighters, its high degree of cooperation among colony members, and its enormous rewards for victory, still kills only when the time is right. When the odds in its favor are unmistakable.

/ / /

Species with coalitionary bonds and variable party size — let us call them the *party-gang* species — are wont to kill adult neighbors.* The number of such species is small, and deaths are in any case rare, so it will take many years to learn how similar and different these species' killing patterns are. But as with infanticide, the underlying formula that links the deliberate killers of the world looks clear, simple, and ignominious. Killing is possible in party-gang species because it is cheap. Power corrupts. Low risk breeds assassins.

Armed with the notion that the likelihood of any action is decided by the economics of individual reproduction, biologists trying to understand evolved behaviors constantly find themselves juggling two things: benefit and cost. Higher benefits and lower costs make the appearance of a behavior more likely. In party-gang species the cost of killing may be low, but what are the benefits? The party-gang species have so much variety in their societies that the benefits of killing neighbors must differ enormously from one instance to the next. If so, then looking for benefits may not be so useful as looking at costs. It may matter little what such a species competes over, or how valuable the prize. Provided that killing is cheap enough, in almost any rivalry killing will pay. At the very least, killing a neighbor reduces competition over resources.

Chimpanzees and spotted hyenas both live in societies that are xenophobic, wandering in small parties, fighting with neighbors. Considering only those aspects, we could imagine both as lawless humans: gunslingers and desperadoes in the Wild West, marooned British schoolboys in *Lord of the Flies*, nihilistic street gangs in South Central Los Angeles.[19] Yet for all their

---

* In this definition of party-gang species, *coalitionary bonds* refers to relationships between same-sex adults used in aggression against others of the same sex, while *parties* refers to temporary groups, which can theoretically range in size from lone individuals to as many as are in the community. Among chimpanzees the average party is less than ten; parties seldom grow beyond twenty.

similarities, their social lives are in many ways polar opposites: mirror images of sex and bonding.

In chimpanzee society, patriarchy rules. Communities persist through a line of father-son relationships. Males are the inheritors of territory. Males conduct the raids and the killing. Males are dominant. Males gain the spoils. But within any spotted hyena community, females rule. Females never leave their natal clans. And in hyena clans, up to eighty strong,[20] the leader is always female — not because the females are larger (they are the same body length as males, on average, but a little heavier, probably because they are better fed), but because they are more determined, more aggressive, and, most important, more united.[21] Females fight harder. Females are dominant. Females gain the spoils.

For chimpanzees, a lost territory means death for adult males, but not for females.[22] Females have more options, more freedom, even if they suffer to take advantage of it. When the K-group community of chimpanzees in Mahale was reduced to a single male in 1979, five fertile females joined the neighboring M-group. At least four infants born subsequently to these females were killed by the M-group males, but after those first killings, later babies were allowed to live.[23] No such luck for males on the losing side. The dominant sex lives and dies by its territory, but the subordinate sex can sometimes emigrate and thereby survive. We have seen the same pattern among humans in primitive wars.

Territorial gains, like territorial losses, have different impacts on males and females. For a male-bonded chimpanzee community, conquered land can include not only a larger foraging area, but also new females who may simply continue to forage in the same area of forest as before the boundaries changed, only now with a different set of defenders. So males of an expanding community can gain females, which means that male chimpanzees should want to expand their territory to the largest area they can defend. Evidence that they do so comes from Gombe, where the territory size of the Kasekela community varied in proportion to the number of adult males.[24] Among hyenas, by contrast, strange females or males in conquered land are not absorbed. They are driven off or killed. But expansion still gives the victors extra hunting grounds, hence extra food for the clan.

Surely the extra food gained by expansionist female hyenas has different value than that of extra females for male chimpanzees — quite as the benefits of human war must differ vastly between agricultural societies with stores worth stealing and some foraging people who fight only over women. Compare any of these examples with pair-bonded wolves, and we find differences again. And most likely the benefits of feuding vary even among different populations of chimpanzees.[25] Does it matter, in other words, *what* the fight achieves? If you're a party-gang species living in rivalry with neighbors, a chance to kill safely tends to pay off for the same underlying reason. It weakens the neighbors. The future can't be foreseen, but whatever it holds, the neighbors will be rivals, armed and dangerous. The stronger you are, the more easily their land can be taken, regardless of the particular benefits the land will bring.

This helps explain why humans are cursed with demonic males. First, why demonic? In other words, why are human males given to vicious, lethal aggression? Thinking only of war, putting aside for the moment rape and battering and murder, the curse stems from our species' own special party-gang traits: coalitionary bonds among males, male dominion over an expandable territory, and variable party size. The combination of these traits means that killing a neighboring male is usually worthwhile, and can often be done safely.

And second, why males? Because males coalesce in parties to defend the territory. It might have been different. If females were the resident sex and formed the coalitionary bonds and defended the territory, humans might still have had Genghis Khans, Alexanders, Caesars, and Hitlers. But they and their favored gods and their trusted soldiers would be women. Hyenas show us that human male violence doesn't stem merely from maleness.

We cannot lay the blame for demonic males on a general tendency among apes for excessively violent behavior any more than we can on an artificial world of culture somehow disconnected from nature at one historical moment in time. Two species-typical behavioral conditions — party-gangs and bonded males — suffice to account for natural selection's ugly legacy, the tendency to look for killing opportunities when hostile neighbors meet.

Nature is like a Russian doll, and each layer of questions we answer reveals another layer to address. Why do chimpanzees and humans form party-gangs when other primates live in stable troops? And why are we male bonded when we could be female bonded or both, or neither?

The easier problem is that of party-gangs, a peculiar style of social behavior that looks entirely explicable by the *cost-of-grouping theory.* This theory states that primate groups might grow infinitely large, except for the restraints imposed by ecological costs. And in habitats where the ecological costs vary seasonally, party size varies accordingly. The notion is well supported by what we know about the party-gang species, because for all of them, parties become bigger when or where more food is available. We see this pattern every year among the chimpanzees in Kibale Forest. In Kibale, the number of trees with fruit oscillates erratically from 0 to 8 percent.[26] And only when many trees have ripe fruit do chimpanzee parties really swell in size. This is a time when the chimpanzees and their observers become equally ecstatic. More food makes bigger parties, which are easier to find, watch, follow, and learn from. That simple pattern, as a matter of fact, is a consistent rule for all the party-gang species, whether chimpanzees, humans, spotted hyenas, or others: more food, bigger parties.[27]

For many other primate species, those with stable troops instead of parties, the cost-of-grouping theory also turns out to work very well. Primate ecologists Charles Janson and Michele Goldsmith cleverly tested the theory by comparing the ratios of travel distance against troop size for various species of primates. When a troop of any species becomes larger, it will have to travel farther during any given day to acquire enough food for all the individuals. So the first expectation of a cost-of-grouping analysis would be that larger troops of any species will travel farther during the day. But Janson and Goldsmith found a far more interesting pattern. They found that for any given increase in troop size, the *proportionate* increase in daily travel distance was much higher for some species than for others. For some species, in other words, the apparent cost of collecting into troops was higher than for others. If the cost-of-grouping theory holds true, then one would expect that those species forced to travel greater daily distances by increased troop size will turn out to be the same species that normally collect themselves into relatively small troops. Grouping is expensive for them, so they prefer small troops. And, yes, the data collected by Janson and Goldsmith fulfill that expectation.[28]

The cost-of-grouping theory appears to explain why some species live in stable troops while other species live in party-gangs. Party-gang species, for ecological reasons, cannot afford to live in permanent troops year-round. They simply happen to possess lifestyles that make grouping very useful at some times and quite costly at others. These are lifestyles centered on eating high-quality but sometimes hard-to-find foods. Foods, perhaps, that pop up seasonally or grow in patches variable in size and density. Foods that are especially nutritious when you can find them, but are often unavailable. Foods that may be abundant one moment and rare the next. Foods like ripe fruits and fat-rich nuts and juicy roots and meats. Foods like the ones that both chimpanzees and humans have evolved to rely on.

So the party-gang patterns of chimpanzees and humans probably come from our being connoisseurs of high-quality foods that are often too scarce to allow friends or allies to forage together regularly without some or all of them starving. If only we were like gorillas and could sit down in a mountain meadow eating leaves all day, we could cheerfully live in stable troops —

just as gorillas do. But our digestive systems didn't evolve to process leaves all day, and thus, while we move in parties when we can — for protection and for the benefits and pleasures of sociability — we still hunt those gourmet foods rare enough that just one extra forager in our party can reduce the gains markedly for everyone else, forcing us all to go distinctly farther every day to fill our bellies or food sacks. And for us connoisseur species, the distances easily become too great, intolerable on our tight energy budget. Rather than starve together, it is better to break up and forage alone, however valuable or pleasant the company.

Like other primates and carnivores, the party-gang species are xenophobic and territorial if economics allow it. But the vicissitudes of hunger and plenty force them into temporary parties, sometimes fusing into larger boisterous groups, other times splitting into smaller groups or even lone individuals. By living as a party-gang species, we gain tremendous flexibility in our hunt for gourmet foods and we grasp the freedom to adapt readily to varying ecological pressures. But as an unhappy price of that freedom, when parties grow smaller they simultaneously become increasingly vulnerable to attack by neighbors who happen to be hanging out for the moment in bigger parties.

And why are those aggressive neighbors almost always male for chimpanzees and humans, but female for spotted hyenas? Partly, perhaps, because female apes don't benefit as much by fighting for food as female spotted hyenas do. You can't aggressively defend a branch of fruit as effectively as you can a chunk of meat. But benefits aside, there's a cost-of-grouping problem that pushes apes toward a male-bonded social system. Adult males walk faster and tire less quickly than adult females with infants because the males don't carry the infants. You see this principle express itself most clearly on long chimpanzee expeditions to favorite food patches. A mixed party may start off together, but mothers carrying their offspring often end up taking a rest halfway up a hill or slowly falling farther behind, so after a twenty-minute walk they arrive at the food patch five minutes after the males. Almost always it is only the females without infants who keep up with the males. So extra travel simply costs less for adult males and for childless females than it does for adult females with infants. This simple fact might on its own explain why males can spend more time together than females. They can afford to. They can afford to travel in larger parties because the extra foraging distance is less expensive for them. And hence, in a classic example of how a seemingly arid ecological issue can ultimately generate major social effects, cost-of-grouping theory suggests that males can bond together merely because they can afford to spend more time together. If this is right, then it also predicts that childless females will be more social than mothers. Indeed they seem to be so. And mothers should form closer bonds with each other during the times and in the places where abundant food means they can spend time together. Again, they do.[29]

Orangutans provide a nice counterpoint. As we noted in the last chapter, females and small males are both faster than big males. And as expected by this theory, both are found in larger parties than big males are — though in both cases the parties are still very small on average.[30]

Around 5 million years ago, our chimpanzee-like ancestor became a woodland ape and spawned a family of descendant species. Around 2 million years ago, the early signs of humanness emerged. Climbing adaptations fell away. Erect bipedalism became more refined. Teeth, mouth, and jaw became smaller. The brain became larger. Changes to mouth and brain continued at a varying pace until modern humans evolved 100,000 to 200,000 years ago. But not until after agriculture began, a mere 8,000 to 10,000 years ago, did human societies begin to unveil their habits clearly.[31] Evidence of real war comes soon after that unveiling. Jericho at 7000 B.C. was a thriving city of 2,000 to 3,000 inhabitants, a center of cultivation within a fertile oasis formed by the River Jordan. Among the population were traders and craftsmen who processed and stored, handled and distributed food and material goods. The stony remnants of that old city today tell us, however, that Jericho was designed as a fortress: surrounded by one continuous wall more than three meters thick and four high, reinforced below by a broad moat cut down into the bedrock another three meters; supplemented above by a lookout tower rising an additional five meters.[32] Written history starts a little later, with scraps of pottery from modern-day Iraq bearing witness to the Sumerian invention of writing in about 3100 B.C. By then, the written record informs us, wars and the patriarchal systems fighting them were in full glory.

The mysterious history before history, the blank slate of knowledge about ourselves before Jericho, has licensed our collective imagination and authorized the creation of primitive Edens for some, forgotten matriarchies for others. It is good to dream, but a sober, waking rationality suggests that if we start with ancestors like chimpanzees and end up with modern humans building walls and fighting platforms, the 5-million-year-long trail to our modern selves was lined, along its full stretch, by a male aggression that structured our ancestors' social lives and technology and minds. A few collected heaps of smashed skulls and projectile tips embedded in bone, the rare but fascinating examples of modern people living in Pleistocene economies, and the occasional vague theoretical glimpses we can gather on occasion of that otherwise deeply hidden 5 million years do not challenge this vision.[35]

And yet, if we take this Dostoyevskian view of our origins, one more mystery immediately presents itself. Surely, if our ancestral males were so demonic in the structure of their daily lives, natural selection should have left deep traces of design in the structure of our own bodies. But humans look so feeble compared to the other apes. So where does the biologist find evidence of adaptation to our postulated demonic past? And if such evidence cannot be found, why not?

9

, , ,

LEGACIES

Sexual selection, the evolutionary process that produces sex differences, has a lot to answer for. Without it, males wouldn't possess dangerous bodily weapons and a mindset that sanctions violence. But males who are better fighters can stop other males from mating, and they mate more successfully themselves. Better fighters tend to have more babies. That's the simple, stupid, selfish logic of sexual selection. So, what about us? Is sexual selection ultimately the reason why men brawl in barrooms, form urban gangs, plot guerrilla attacks, and go to war? Has it indeed designed men to be especially aggressive?

Until we have carefully examined the evidence, our answer should be: not necessarily. Because the social, environmental, genetic, and historical circumstances for any single species are so extremely complex, we can't assume a priori that sexual selection has acted in any particular way for any single species. Among the 10 million or more animal species on earth, you can find interesting exceptions to almost every rule. On the one hand, you can find species like spotted hyenas, where such extraordinary ferocity has evolved among females that it outshines even the stark sexual aggression shown by males. And on the other, you will discover the pacifists.

Among primates, the most extreme pacifist is a little-known South American monkey whose ordinary life might warm the hearts of optimists everywhere. The muriqui is the largest monkey in South America, an elegant tree dweller with a tail so muscular and lithe that it acts as a fifth limb. Confined to a few patches of dwindling forest nestled among the coffee estates of southeastern Brazil and extremely rare in captivity, it has only occasionally been filmed or photographed. Primatologists knew nothing about the species' social life until the 1980s, when it was discovered that several adults of both sexes travel together in small groups to search out patches of fresh fruits and leaves.

In many ways their lives are very ordinary — but not when it comes to sex. Females can't be bullied, and they choose their mates at will. They often mate in full view of several males, in flagrant copulations that last for an average of six minutes and up to eighteen. For most other primates, such open sexual activity would create a flurry of agitated, aggressive attempts by males to stop the female from mating with others. Not for muriquis. Muriqui males watch calmly, and sometimes they casually take turns. Up to four males may sit in line along a branch, patiently watching all that sex, and waiting their moment.[1] While doing so they never show the slightest emotion — no aggressive displays at all.

Sexual jealousy is such an ordinary and predictable emotion in our own species that people have trouble imagining a species in which it seems utterly absent. Perhaps, you might think, those waiting males are really burning up inside. Maybe in front of the female they simply don't dare express their feelings of competition. But we know that the muriqui males are not merely temporarily inhibited by that particular situation. Fact is, they never show any signs of concern about hierarchy, about who gets there — anywhere — first. Muriquis have no alpha male, no social ranks, no dominance relationships, not even any status-seeking. They are all equal. Males show aggression only when they meet a different group — and so far, even that has been mild. One for all and all for one is their credo. If any male primate looks to be egoless, it's the muriqui.[2]

In an appropriate anatomical parallel to their behavior, the bodies of male muriquis show little evidence of selection for aggression (and the same is true for females). Males are the same size and weight as females, and males' canine teeth are about the same length as the females'.[3] In short, their bodies, like their minds, do not seem to be especially designed for fighting. What a contrast to chimpanzees! In that species males are bigger than females, they have longer canine teeth that serve as weapons, and they express their violent tendencies in fights and coercion and raids and bullying. Male chimpanzees are relentlessly driven to achieve higher status, whether in captivity or in the wild. They never have social relationships with other males as calm as those of the muriqui. Nor have gorillas or orangutans. We can confidently conclude that something in the genes of these male apes — chimpanzees, gorillas, orangutans — gives them fighting bodies and leads them to compete with each other regardless of circumstance. But that mysterious something turns out to be missing in the muriqui.

This comparison points to a general truth. Just as individuals of the same species will certainly vary in temperament (partly as a result of their upbringing), so too, when we compare the typical or average temperament of a species, we find a wide variation between species.[4] By showing us just how much variation there can be, muriquis expand our concept of how sexual selection works. Instead of designing males as fighters, always and inevitably, sexual selection can do quite the reverse. It can produce a species with males who are friendly and harmless, whose competition with each other is expressed not through fighting but through winning the race to find the female (as happens in some fish), or through producing sperm of higher quality and larger numbers than rivals (as appears to be the case with muriqui). Why sexual selection produces different results in the evolution of different species isn't always easy to discern; and in the case of the muriqui, the logic of their sexual selection has not been fully worked out, but the facts themselves are clear. Muriquis show us that male primates can evolve into beings with uncompetitive, or at least unaggressive, temperaments.*

The muriqui offers an enticing image. For those who believe that humans evolved as a pacific species whose modern-day violence can be blamed on something other than our evolutionary past, it's nice to find a real-life example — albeit one that is only distantly related to us. The notion of an ancestry of pacifism is widespread today, an essential component of the cultural determinist legacy. It is probably not an exaggeration to say that the "culture only" theory of human violence is conventional wisdom. This potent assumption, for example, led an international group of twenty distinguished scientists to sign a formal declaration in 1987 pronouncing warfare to be "a peculiarly human phenomenon," one that "does not occur in other animals," a strangely destructive sort of activity scientifically proved to be "a product of culture," having only a minor "biological connection . . . primarily through language." Known as the Seville Statement on Violence and prepared on behalf of UNESCO, this declaration placed the ultimate seal of approval on the position that "biology does not condemn humanity to war," and that therefore humanity can soar to freedom, having broken away from "the bondage of biological pessimism." Quite a statement. Its motives were clearly on the side of the angels. But liking the idea doesn't make it right.

Much of the logic that underlies it is certainly wrong, for the simple reason that it falls into that century-old trap, Francis

* There is always competition between the genetic interests of different individuals. But males may compete for reproductive success in ways that don't require competitive temperaments. Among other things, muriqui compete through a system described by biologists as *sperm competition*. Males have enormous testes, produce copious amounts of semen, and often remove each other's seminal plugs from a female before mating.

Galton's false dichotomy, Galton's Error. Particularly when important issues are at stake, this attractive oversimplification — that species' characteristics must come from either nature or nurture but not from both — lures even brilliant people into a wrong conclusion. We know that any person's tendency to be violent is influenced by all sorts of circumstances, from family history and alcohol use to cultural norms, economic conditions, political and historical contexts, and so on. An individual's violent behavior is influenced in any number of different ways by the environment — by nurture. Those environmental influences are important. Who could disagree? But then Galton's Error hits home. Stumbling over the buried idea that characteristics must be produced by either nature or nurture, many people feel forced to choose. Since nurture obviously has effects, nature must be denied. Thus, Galton's Error leads to the false conclusion that aggression cannot come from nature. And the essence of this fallacy sometimes settles down into a simple mantra. Behavior that varies during an individual's lifetime does so because tradition or accident alters its course. So whatever varies cannot come from the genes.

But it can, of course! In countless instances biologists have watched animals change their behavior to suit their purposes. Indeed, the whole logic of evolution would indicate that animals use their intelligence to serve evolutionarily appropriate goals. Why otherwise would problem solving and learning (and the variable behavior that these abilities create) ever have evolved? Complex animals have complex mental and emotional systems underlying their behavior. These systems have evolved, and they are subject in turn to genetic variation. Inherited temperaments in different environments can express all sorts of different behaviors. Even in a single species we see bursts of creativity and vastly different approaches to solving similar problems, but the underlying psychology remains inescapably subject to evolutionary forces.

So there is no particular reason to think that human aggression is all cultural, or that our ancestors were as pacific as muriquis. The only way to find out whether sexual selection has shaped human males for aggression is to leave the theory and go back to the evidence. There are two places to look for an answer. We can look at our bodies, and we can think about our minds. The easier part is our bodies.

A biologist from Mars looking at a preserved human male laid out on a slab might find it hard to imagine our species as dangerous. Lined up next to male specimens from the other apes, or from virtually any other mammal species, human males don't look as if they are designed to fight at all. They are rather slender, their bones are light, and they appear to have no bodily weapons. People don't think of humans in the same way that they think of dangerous animals. That first impression is misleading, however. Humans are indeed designed to fight, although in a different way from most of the other primates.

Here's one clue. Men are a little larger and more heavily muscled than women. For other primate species, larger male size links strongly to male aggression. But with humans, that apparent evidence seems to conflict with the absence of fighting canine teeth. Could it be that humans break the general rule linking larger males to an evolved design for aggression?[5]

Consider our teeth. The upper canines of most primates are longer and sharper than any other tooth. These long teeth are obvious weapons, bright daggers ground to a razor-sharp edge against a special honing surface on a premolar tooth in the lower jaw. Baboons, for example, have canines five to six centimeters long. Male baboons trying to impress each other grind their canines noisily, occasionally showing off their teeth in huge, gaping yawns. When male baboons make those display yawns, they are acting like cowboys twirling their revolvers.

By comparison, human canines seem tiny. They barely extend beyond the other teeth, and in males they are no longer than in females. Those canines may help us bite an apple, we love to imagine them elongated for the Halloween scare, and we unconsciously display them when we sneer, but our canines virtually never help us fight. In fact, the fossil record indicates that ever since the transition from rainforest ape to woodland ape, our ancestors' canines have been markedly smaller than they are in chimpanzees. In the woodlands, those teeth quickly became muriqui-like in appearance — one reason why some people wonder if woodland apes were as pacific as modern muriquis are.

But we should not allow ourselves to become misled by the evidence of canine teeth. The importance of a species' canines depends entirely on how it fights. Chimpanzees, for example, have distinctly smaller canines than baboons. Does this worry chimpanzees when they fight the formidable, saber-toothed baboons?

On an East African dry season afternoon in 1972, I watched a chimpanzee fight a baboon. Hugo Chimpanzee came to a palm tree where Stumptail Baboon was already ensconced, eating fruits. It was an especially harsh part of the season, when fruit patches were few and palm fruits were prized. Hugo spent several minutes looking at the palm crown from various angles. Eventually he appeared to satisfy himself that there was enough fruit to justify the climb, and he started upward. After a minute he arrived on the opposite side from Stumptail. Hugo caught his breath, then moved carefully around the crown to challenge Stumptail for possession of the only feeding site on the tree. I was nervous for Hugo. At 40 kilograms, he was about twice as big as Stumptail, but weight would count for little in their precarious perch. Moreover, Stumptail had enormously more impressive teeth and knew how to use them. As Hugo approached, Stumptail reared, bared his fangs, and threatened hard — but before he could close to biting range, Hugo swung his arm in a wide arc and punched Stumptail in the belly. Stumptail crumpled forward, looking sick. Moving like a prizefighter, Hugo quickly landed a second punch on Stumptail's chin, snapping the baboon's head backwards. That was it. Stumptail retreated and made himself scarce for the rest of the day. And Hugo, taking his place among the delicious palm fruits, ate for a peaceful half hour.

Apes can fight with their fists because they have adapted to hanging from their arms, which means that their arms can swing all around their shoulders, the shoulder joint being a flexible multidirectional joint. So chimpanzees and gorillas often hit with their fists when they fight, and they can keep most canine-flashing opponents at bay because their arms are long. If chimpanzees and gorillas find punching effective, then surely the woodland apes, who were standing up high on hind legs, would have fought even better with their arms.

Fists can also grasp invented weapons. Chimpanzees today are close to using hand-held weapons. Throughout the continent, wild chimpanzees will tear off and throw great branches when they are angry or threatened, or they will pick up and throw rocks. Humphrey, when he was the alpha male at Gombe, almost killed me once by sending a melon-size rock whistling less than half a meter from my head. They also hit with big sticks. A celebrated film taken in Guinea shows wild chimpanzees pounding meter-long clubs down on the back of a leopard. (Scientists were able to get that film because the leopard was a stuffed one, placed there by a curious researcher. The chimpanzees were lucky to find a leopard so slow to fight back.) Chimpanzees in West Africa already have a primitive stone tool technology, and there could well be a community of chimpanzees today, waiting to be discovered, who are already using heavy sticks as clubs against each other. Certainly we can reasonably imagine that the woodland apes did some of these things.

So the modern rainforest apes fight with their fists and arms. Our woodland ape ancestors most probably were fistfighters, too. Our legacy from this piece of evolutionary history is much the same as that inherited by kangaroos. Kangaroos given gloves in circus sideshows are famous for their success at boxing humans. In the wild also, kangaroos fight with their arms and hands. In some species, a male uses his arms to hold an opponent at the proper distance for a well-placed kick. In others, he grapples to get into biting position. In both instances, though, sexual selection — because the males are fighting for access to reproductive opportunity — has favored male upper-body strength. Of course, upper-body strength would not be so useful until reproduction is possible; and kangaroo males and females have the same size shoulders and arms until puberty. But at puberty, suddenly, the males develop larger and stronger arms and shoulders. Across different species of kangaroos, development of the arms exquisitely reflects their role as weapons.[6]

The same is true of humans. The shoulders of boys and girls are equally broad until adolescence; but at puberty, shoulder cartilage cells respond to testosterone, the male sex hormone newly produced by the testes, by growing. (In an equivalent way, pubertal girls get wider hips when their hip cartilage cells respond to estrogen, the female sex hormone.) The result is a sudden acceleration of shoulder width for boys around the age of fourteen, associated with relative enlargement of the upper arm muscles.[7] In other words, the shoulders and arms of male humans — like the neck muscles of a red deer, the clasping hands of a xenopus frog, or the canine teeth of many other primates — look like the result of sexual selection for fighting. All these examples of male weaponry respond to testosterone by growing. They are specialized features that enlarge for the specific purpose of promoting fighting ability in competition against other males. Small wonder, then, that men show off to each other before fights by hunching their shoulders, expanding their arm muscles, and otherwise displaying their upper-body strength. Or that male kangaroos flex their biceps at each other in attempts to defeat their rivals without a fight.

If the bipedal woodland apes fought with fists and sometimes with weapons, those species should have had especially broad shoulders and well-muscled arms, like modern men. We haven't enough fossils yet to know if that's true. Indeed, it is not yet absolutely certain that male woodland apes were larger than females, though most of the current fossil evidence suggests so.

If they were, we can confidently imagine that the males were designed for aggression. Perhaps the early development of club-style weapons might also explain why the skulls of our ancestors became strikingly thicker, particularly with *Homo erectus* at 1.6 to 1.8 million years ago.[8] That's a guess, but it's clear in any case that our present bodies carry the same legacy of sexual selection as other mammals whose males fight with their upper bodies. The broad shoulders and powerful, arching torso we so admire in Michelangelo's *David* are the human equivalent of antlers. The mark of Cain appears in our shoulders and arms, not in our teeth.

What about our minds? Has sexual selection shaped our psyches also, in order to make us better fighters? Can sexual selection explain why men are so quick to bristle at insults, and, under the right circumstances, will readily kill? Can our evolutionary past account for modern war?

Inquiry about mental processes is difficult enough when we deal just with humans. Comparison with other species is harder still. The supposed problem is that animals fight with their hearts, so people say, whereas humans fight with their minds. Animal aggression is supposed to happen by instinct, or by emotion, and without reason. Wave a red rag in a bull's face and the bull charges thoughtlessly — that's the model. Human wars, on the other hand, seem to emerge, so Karl von Clausewitz declared, as "the continuation of policy with the admixture of other means."[9] According to historian Michael Howard, human wars "begin with conscious and reasoned decisions based on the calculation, made by both parties, that they can achieve more by going to war than by remaining at peace."[10] The principle seems as true for the measured deliberations on the top floor of the Pentagon as for the whispered councils among the Yanomamö, and it suggests a wholly different set of psychological processes from the supposedly rigid, instinctual, emotional drives of animals. The fact that we possess consciousness and reasoning ability, this theory says, takes us across a chasm into a new world, where the old instincts are no longer important. If there is no connection between these two systems, the rules for each cannot be the same. In other words, aggression based on "conscious and reasoned decisions" can no longer be explained in terms of such evolutionary forces as sexual selection.

The argument sounds fair enough, but it depends on oversimplified thinking, a false distinction between animals acting by emotion (or instinct) and humans acting by reason. Animal behavior is not purely emotional. Nor is human decision-making purely rational. In both cases, the event is a mixture. And new evidence suggests that even though we humans reason much more (analyze past and present context, consider a potential future, and so on) than nonhuman animals, our essential process for making a decision still relies on emotion.

True, a few behaviors of even such intelligent animals as large mammals are strikingly hard-wired. The infants of spotted hyenas give a good example of the extreme. Spotted hyena babies are very charming, softly furred in black, with the typically cute look of the newly born. Hyenas normally bear twins[11] in a dark den with only the mother present. Unlike any other carnivore, including even their close cousins the striped hyenas, these appealing cubs emerge with fully functional front teeth, including strong, gripping incisors and long, puncturing canines.[12] Their eyes are open, their necks and jaws are strong.

Why the precocial teeth? Why the coordinated heads? For killing each other. Fratricide is routine. Experiments in captivity show that the first clear inclination of the newly born spotted hyena is to bite, then shake the head with muscles unnaturally strong for one so young. In captivity they'll bite at anything, even a cloth. Down in the gloom of the den there are no cloths, but there is a twin sibling, born within an hour of the first and destined to be attacked quickly, even sometimes before it has left its amniotic sac. Sometimes the secondborn may fight back well enough to win. But whoever wins, the weaker cub will often die, most likely from starvation, loser in a brutal competition for its mother's milk. Biologist Laurence Frank and his colleagues estimate that in Kenya's Masai Mara Game Reserve a quarter of all hyena babies are killed by their twins.[13] The violent baby vividly reminds us how starkly aggressive behavior can be shaped by natural selection.*

The hyena baby shows an extreme version of instinctual aggression among mammals. If that is what constitutes aggres-

---

* Natural selection can also shape behavior and morphology to be defensive, of course. Both sexes of hyenas have thick pads on their backs and shoulders, where wounds can accumulate dramatically in subordinates. But the design for defense perhaps reaches its extreme in the appearance of female hyenas. Female hyenas have been known since antiquity to have genitals that look amazingly like those of males. Their clitoris looks just like a penis — in size, in shape, and in the location of the urinary canal. Likewise, their labia are constructed in such a way as to be indistinguishable from the male's scrotum. And, most remarkably, these pseudo-scrota even contain fatty bodies that quite resemble testes. In striking contrast to most evolutionary explanations, this sexual mimicry, so complete, so rare, is normally explained as a chance result of another process — the selection for females to win fights for food.

The accepted wisdom is that the hyena's sexual mimicry results from high levels of androgens in the fetal environment. According to this idea, the benefits of aggressiveness have led to the evolutionary development of high androgen levels among females — and high androgen levels have accidentally, as a meaningless byproduct, masculinized their female infants' physical appearance. Certainly, there is a large advantage to females' being high-ranking: the highest-ranking female hyenas have two and a half times the reproductive success of the lowest. And indeed, researchers have thoroughly studied the masculinizing effects of a mother's high androgen levels on fetus brain and body during critical periods of pregnancy, at least for humans and a few other species — cattle, possums, mice, guinea pigs, rats, hamsters, beagles, and monkeys. Among humans, the effects of an abnormal hormonal environment during pregnancy have been documented in cases where women have taken, for medical reasons, synthetic and natural sex hormones, and in instances where clinical abnormalities have naturally altered the ordinary hormonal balance. When the fetus is biologically female and the mother for some reason expresses high male hormone levels, the female fetus can develop what looks to be male external genitalia — the

sion by emotion, however, it illuminates by being so unusual. Even for spotted hyenas aggression is normally far more Machiavellian. The death of an alpha female, for instance, will elicit fighting only among those females who have sufficient allies to provide a good chance for winning the top slot, with dominance battles appearing carefully planned to the aggressor's advantage.

clitoris enlarges, the labia fuse — as well as a whole host of traditionally "masculine" behaviors in the growing child. The result is a clinical pseudohermaphrodism, ordinarily treated by surgery and hormone therapy. Comparing girls who were fetally androgenized with their normal sisters suggests several behavioral effects of "masculinizing": significantly higher levels of energy expenditure during play, preference for boys as playmates, more initiation of fighting, fewer fantasies about motherhood, much less interest in dolls, higher aversion to infant care, preference of functional over attractive clothes, no interest in jewelry or makeup and hair-styling.

So it makes some good sense that hyena mothers, with their abnormally high levels of male sex hormones during pregnancy, are producing female offspring who look male. But there may be a second factor at work here sustaining this strange condition, because female hyenas take their male mimicry to such extraordinary lengths. Apart from anything else, females not only urinate through their clitorises, they also give birth through them! This procedure is so difficult that the clitoris has to be torn by the passage of the baby, causing first-time mothers to have a high proportion of stillbirths and a significant risk of maternal mortality from abnormal labor, perhaps more than 18 percent. This does not sound like an accidental byproduct of masculinization, because selection should have acted to reduce these costs. Even copulation is made difficult, since it involves the male inserting his penis right into the female's clitoris. (At least it's made easier by special developmental events, including enlargement of the external meatus, the passage that receives the penis, increasing elasticity.) There is no reason in principle why the androgens of the aggressive female should masculinize the genitals so well: Selection could easily have made those tissues insensitive to the male sex hormones. The near-perfect mimicry of male hyenas by females looks like, rather than an accidental outcome of something else, a work of "design" — by which we mean the unconscious design of natural selection.

No reason for this amazing sexual mimicry has yet been advanced

With chimpanzees an even more extended and complex analysis takes place. Male rivals wait strategically for the right moment; when it comes, they try to choose the best tactics. Aggression among males within a chimpanzee community happens most obviously at "election time," during those particular moments when suddenly the old hierarchy is being challenged. Such times occur particularly when a young, low-ranking male whose physical and political power is growing develops a disrespectful attitude toward established authority, typically expressed as a refusal to grovel before a higher-ranking senior. Un-

---

that sounds convincing. But let us put two facts together. First, the most perfect mimicry occurs in the newborn. Second, female newborns viciously attack other females — but not males. Now imagine that you are a female newborn alone with your sibling in the deep narrow passages of the den, where your mother can't help you and your survival depends on your relations with your sibling. Your sibling is a female. And females attack females. If your sibling treats you as a male, her attacks may be reduced. In the war between the babies, every advantage can help. The perfection of sexual mimicry could therefore be an adaptation for protection from intrasexual aggression. It is, we hypothesize, a form of protective camouflage, whereby the female defends herself from other females of her own species by covering herself with the body of a male. The same principle could even work in interclan encounters occurring at night between individuals that don't know each other well. We accordingly interpret the high costs of motherhood that come from having masculinized genitals as the selected result of a beneficial system of defense, not merely a chance outcome of elevated androgens.

So the maintenance of this elaborate sexual mimicry in such precise detail could well be part of an evolved adaptation. At the same time, the high levels of masculine hormones clearly promote the mimicry and simultaneously help sustain the aggressiveness of the females. That these demonic females elevate male sex hormones rather than female sex hormones as a way of increasing or sharpening their aggressiveness implies that they are somehow "masculinizing" themselves for violent purposes. And their reliance on male hormones to stimulate aggressiveness reminds us that violent aggression is — across species — a more common competitive strategy for males than for females. Even for a demonic female species, male hormones are associated with aggression.[14]

settled dominance relations between upstart junior and established senior can spread to their alliance networks and bring violence to the whole community, much as, among humans, a struggle for power in the criminal underworld can terrorize a city. During these power struggles within the chimpanzee community, the overall rates of attack increase twofold or more.[15] The males' apparent emotional motivations seem terribly familiar to the human observer. But when opponents meet, their attacks do not appear to be the wild productions of blind rage. Attacks can be deliberately planned, and sometimes they look surprisingly well considered.

Jane Goodall has described how Mike rose in the male hierarchy at Kasekela. "Once, for example, as a group of six adult males groomed about 10 meters away, Mike, after watching them for six minutes, got up and moved toward my tent. His hair was sleek and he showed no signs of any visible tension. He picked up two empty cans and, carrying them by their handles, one in each hand, walked (upright) back to his previous place, sat, and stared at the other males, who at that time were all higher ranking than himself. They were still grooming quietly and had paid no attention to him. After a moment Mike began to rock almost imperceptibly from side to side, his hair very slightly erect. The other males continued to ignore him. Gradually Mike rocked more vigorously, his hair became fully erect, and uttering pant-hoots he suddenly charged directly toward his superiors, hitting the cans ahead of him. The other males fled."[16] Mike had found a new technology for making dominance displays. He had gone to Goodall's tent, taken the technology, walked back to where the members of the old male hierarchy sat, and very smartly struck and rolled the shiny and noisy metal cans in front of his own charging body, thereby demonstrating effectively enough his own rising status as a contender. No blind instinct here.

One doesn't have to be particularly generous to attribute at least a limited sense of reason to chimpanzees. Their aggression routinely looks rational in the specific sense that it is guided by a complex assessment of the immediate context. Mike acted, for example, as if he were thinking how best to achieve his aim. However, this doesn't mean he acted without emotion. Reason showed him how to dominate the high-ranking males. But emotion made him want to. Here's a nice parallel to Galton's Error. When we think of the ultimate influences on behavior, we should think of nature and nurture as complementary, not as mutually exclusive alternatives. In the same way, when we think about the mental processes behind animal behavior, we should think of emotion and reason as complementary, not alternative.

And the same is true when we think about human behavior, according to a theory fully elaborated by Antonio Damasio, head of the department of neurology at the University of Iowa College of Medicine. Conventional wisdom claims that people solve problems and make decisions by evaluating several possible solutions or actions and then choosing one of them — ideally, the best. People normally describe this process as "rational," in the same sense that war historian Michael Howard uses the word "reasoned" to describe the human calculations over whether or not to go to war. That model seems appropriate at first.

But Damasio's research provides another way to understand how people decide. Damasio has studied brain-damaged patients who are fully normal in most ways but inhibited or mentally paralyzed in one particular way: They can't make decisions. Combining his research on living patients with an intensive review of the historical record (including a computer-assisted reconstruction of the nineteenth-century's most spectacular brain damage case, that of railroad worker Phineas Gage, who, after an explosion sent a three-and-a-half-foot-long iron rod sailing up through his cheek and out the top of his skull, remained fully conscious and intellectually intact, well and personable enough to be exhibited as a freak in Barnum's American Museum in New York), Damasio found that all of these patients shared one thing in common. They had all been injured or damaged at a particular location, the ventromedial part of the prefrontal cortex. Brain damage in this region leads to two main results: first, a general lack of initiative; second, a strange emotional vacuum, so that the sufferer is dispassionate and uninvolved in the surrounding world, no longer caring about life.

Damasio and his team worked closely with one such patient, a man called Elliot. Elliot's intellectual ability, social sensitivity, and moral sense were examined with a battery of tests on which he performed quite well. In most circumstances, Elliot communicated and interacted as any normal person would. He had a cool wit, a good sense of humor. He was aware of cause and effect. He could devise appropriate solutions to hypothetical social problems. And he described easily the consequences of theoretical solutions to hypothetical problems. But when asked actually to solve a problem, all he could do was analyze again and again the several options. He could never choose one. This sort of intellectual stasis — the inability to choose a course of action — deeply affected his daily life. After sustaining brain damage, Elliot wouldn't even get up in the morning without being prompted, and at work his initiative was utterly absent. He simply persisted in any given task, never deciding that it was time to move on to the next problem. Though he could think as well as ever, he had lost the capacity to decide.

Such patients cannot decide, thinks Damasio, because their brains are unable to connect an emotional value to the intellectual menu of possible options. Without being able to feel which solution they like, they have no way to choose. These patients seem to demonstrate that pure reason is inadequate for reaching a decision, a hypothesis that Damasio applies to all of us. Reason generates the list of possibilities. Emotion chooses from that list.[17]

Humans, of course, possess vastly superior powers of reasoning compared to chimpanzees. And in a similar way, chimpanzees are cognitively superior to most other animals. At the same time, however, according to Damasio's model, these differences in reasoning ability between species still may not change the essential structure of decision-making. An individual of a cleverer species may generate more mental options, manipulate them more consciously, and be able to follow the logical consequences further; but having done so, the individual is still left with a set of options to choose from. Human or chimpanzee, the problem-solver's final mental act in the decision-making process depends on emotion. Option A will lead to Outcome 1. Option B will lead to Outcome 2. For each outcome the brain must now generate an image, attach an emotional quality to the image, and compare how the different emotions feel. *I like the feel of the image produced by Outcome 1. But I like the feel of the image produced by Outcome 2 even more!*

People have always accepted that animals act from emotions; humans, says Damasio, can never act without them. Suddenly the apparent chasm between the mental processes of chimpanzees and our species is reduced to a comprehensible difference. Humans can reason better, but reason and emotion are linked in parallel ways for both chimpanzees and humans. For both species, emotion sits in the driver's seat, and reason (or calculation) paves the road.

We are now ready to ask what causes aggression. If emotion is the ultimate arbiter of action for both species, then what kinds of emotions underlie violence for both? Clearly there are many. But one stands out. From the raids of chimpanzees at Gombe to wars among human nations, the same emotion looks extraordinarily important, one that we take for granted and describe most simply but that nonetheless takes us deeply back to our animal origins: pride.

Male chimpanzees compete much more aggressively for dominance than females do. If a lower-ranking male refuses to acknowledge his superior with one of the appropriate conventions, such as a soft grunt, the superior will become predictably angry. But females can let such insults pass. Females are certainly capable of being aggressive to each other, and they can be as politically adept as males in using coalitions to achieve a goal.[18] But female chimpanzees act as if they just don't care about their status as much as males do.

By contrast, we exaggerate only barely in saying that a male chimpanzee in his prime organizes his whole life around issues of rank. His attempts to achieve and then maintain alpha status are cunning, persistent, energetic, and time-consuming. They affect whom he travels with, whom he grooms, where he glances, how often he scratches, where he goes, and what time he gets up in the morning. (Nervous alpha males get up early, and often wake others with their overeager charging displays.) And all these behaviors come not from a drive to be violent for its own sake, but from a set of emotions that, when people show them, are labeled "pride" or, more negatively, "arrogance."

The male chimpanzee behaves as if he is quite driven to reach the top of the community heap. But once he has been accepted as the alpha (in other words, once his authority is established to the point where it is no longer challenged), his tendency for violence falls dramatically. Personality differences, and differences in the number and skill and effectiveness of his challengers, produce variation in how completely he relaxes. But once males have reached the top, they can become benign leaders as easily as they earlier became irritated challengers. What most male chimpanzees strive for is being on top, the one position where they will never have to grovel. It is the difficulty of getting there that induces aggression.[19]

Eighteenth-century Englishmen used less dramatic tactics than wild chimpanzees, but that acute observer Samuel Johnson thought rank concerns were as pervasive: "No two people can be half an hour together, but one shall acquire an evident superiority over the other." Pride obviously serves as a stimulus for much interpersonal aggression in humans, and we can hypothesize confidently that this emotion evolved during countless generations in which males who achieved high status were able to turn their social success into extra reproduction. Male pride, the source of many a conflict, is reasonably seen as a mental equivalent of broad shoulders. Pride is another legacy of sexual selection.

And can it account for war? The immediate causes of wars are as varied as the interests and policies of those who launch them, but deeper analysis leads to a consistent conclusion: Wars tend to be rooted in competition for status.

The first great multistate war, the Peloponnesian War, which ravaged Greece from 431 to 404 B.C. and led to the total defeat of Athens, was studied for years by the first great historian, Thucydides, who eventually felt he understood its cause: "What made war inevitable was the growth of Athenian power and the fear this caused in Sparta."[20] Sparta and Athens were old allies who had been united against the tyranny of Persia. By virtue of her armies, Sparta had dominated the Peloponnesian League. In fact, the League was at one time known as "Sparta and her allies." But during the fourth century B.C., as a consequence of an expanding Athenian naval power, Athens came to develop her own empire. The rivalry between Sparta and Athens broke out into war, fed by border clashes and conflicts over trade, prospects of booty, individual acts of treason or glory-seeking, and by all the complex divided loyalties and personal ambitions that mark any war. But the essential dynamic, according to Thucydides, was that Sparta watched the growth of Athenian power, feared the outcome, and decided to counter the threat.[21] Michael Howard argues that the same logic applies throughout history, from the Peloponnesian War to the World Wars of the twentieth century. Men fight, he says, "neither because they are aggressive nor because they are acquisitive animals, but because they are reasoning ones: because they discern, or believe that they can discern, dangers before they become immediate, the possibility of threats before they are made."[22]

We could well substitute for Sparta and Athens the names of two male chimpanzees in the same community, one rising in power, the other anxious to keep his higher status.

Even if Athens's power had exceeded Sparta's, the future would still be uncertain. Athens might still have remained benign or been devastated by natural disaster or conquered by another enemy. But Sparta knew, of course, that Athens would likely take advantage of her power, because, apparently, Greek city-states behave the way human or chimpanzee males behave. So it is easy to see the Peloponnesian War as having begun as a consequence of competition between two prideful city-states, ruled by prideful men concerned — like two tough guys squared off for a fight in a tavern — not rationally but emotionally about who is the biggest and best. Which city-state is number one? Who is the real superpower of the Peloponnesus? Pride, the emotional complex driving status competition, may remind us of the Peloponnesian War, or it may make us think of contemporary street gang clashes, such as those recalled by Sanyika Shakur from the early days of Crips and Bloods in South Central Los Angeles. "Our war, like most gang wars, was not fought for territory or any specific goal other than the destruction of individuals, of human beings. The idea was to drop enough bodies, cause enough terror and suffering so that they'd come to their senses and realize that we were the wrong set to fuck with. Their goal, I'm sure, was the same."[23]

Sparta's seemingly rational fear, laid in place by evolved systems of thinking and feeling, was based on calculated guesswork about Athenian intentions — but also on the unexamined feeling that it's always worth being on top, an emotion that has evolved for good reasons. We can doubtless overcome this feeling, but the bias in favor of it is powerful. We get into fights or lust for imperial dominion over another nation for reasons of pride.[24]

Men come physically armed for aggression, and they look emotionally primed to pursue high status. But the same could be said of a solitary species like the orangutan, whereas humans, like chimpanzees, are extremely social. In many ways the most interesting questions about human male temperament concern the legacy of social aggression.

In 1960 three close friends from New York went to college in

North Carolina on basketball scholarships. All three were first-rate players. They would be separated because two of them had chosen one college and the third another, but they were best buddies. They would still be in the same state. So the three knew they would still keep up their old strong ties. What these young men didn't realize at the time, however, was that their two colleges, Duke and the University of North Carolina, maintained a bitter rivalry. Art Heyman went to Duke; Larry Brown and Doug Moe went to North Carolina. And the resulting competition broke up their friendship. By the time they played each other in a freshman game, they were not merely former friends, they were serious enemies. Moe spat at his erstwhile friend Heyman. And by next season, after Brown and Heyman bounced off each other during a game, they squared off in a fight so earnest it took ten police officers to pull them apart.[25]

The conflict between individual friendships and group loyalty is the stuff of a thousand plays and books and operas and histories. Groups command extraordinary devotion, even groups as disconnected from our evolution in kin-based communities as basketball teams. But why should they? Is the lure of the group a result of rational deliberation, or is it the response of an old ape brain?

Social psychology doesn't ask that question, but it has nevertheless shown that group loyalty and hostility emerge with a ridiculously predictable ease. The seminal experiment was performed at a summer camp near Robber's Cave, Oklahoma.[26] There, during the 1950s, twenty-two middle-class, white, Protestant, well-adjusted, eleven-year-old boys were invited to camp. The psychologists working with the camp then split the boys into two groups that were kept apart. Friends were separated as much as possible. The experimenters' aim: To find out how easily group hostility would emerge.

It took a week for each group to give themselves an identity, a leader, and a culture. One group called themselves the Rattlers, and they prided themselves on being tough, refusing, for instance, to complain about injuries. The other group, the

Eagles, concentrated on vilifying homesickness. Finally, a grand tournament was announced. For five days the Rattlers and the Eagles would compete for trophies.

It started well enough, with a baseball game played fairly. The Eagles lost. But that night, in a sneak raid, the Eagles burned the Rattlers' flag. Next day, the Rattlers' leader started a fight by challenging the Eagles' leader to admit to burning the flag. When other kids joined in, experimenters felt they had to intervene. The intervention stopped the fight for the time being, but then, hours later, it escalated. That night the Rattlers raided the Eagles; the Eagles retaliated with a raid of their own. This time the boys were fighting with sticks and bats, and they were prepared to arm themselves with stones. But then the Rattlers stole the Eagles' trophies, and when the Eagles negotiated to get them back, the Rattlers insisted on humiliating the Eagles by making them crawl on their bellies. Finally, the experimenters, seeing that things were getting out of hand, restored a degree of friendly relations among the boys by setting goals that could be accomplished only by the two groups working together.

The Robber's Cave experiment observed children in the woods. But the same kind of experiment works as well with adults in industry, and it has been repeated convincingly in dozens of other contexts.[27] People quickly form groups, favor those in their own group, and are ready to be aggressive to outsiders. We all abhor such biases. They lead to all sorts of "isms" — racism, sexism, ethnocentrism, and so on. But we all fall prey to them with an appalling ease.[28]

The process begins, say social psychologists, with people categorizing, mentally putting people into coarse and general classes that ultimately boil down to Us and Them. Next, people start to discriminate, favoring Us over Them, even when the basis for assortment is totally meaningless — for example, whether a person overestimates or underestimates the number of dots on a screen. Finally, they stereotype. They say nice things about Us and nasty things about Them.[29]

The temperamental complex involved is what we call the

ingroup-outgroup bias. The ingroup-outgroup bias is often ethnocentric, which means that the ingroup and the outgroup are perceived as different races or ethnicities, but it can appear with equal ease around other categories, such as religion, sex, age, or sports team. In striking contrast to many or most processes they describe, social psychologists describe this complex as universal and ineradicable.[30] Taken to an extreme, ingroup-outgroup bias effectively dehumanizes Them, which means that moral law does not apply to Them and that therefore even ordinary and very moral people can do the most appalling things with a clear conscience. During the first fifty years of the Spanish conquest of the New World, the Spaniards regarded the Indians as subhuman. They treated their dogs better, according to a contemporary witness, "terrorizing, afflicting, torturing, and destroying the native peoples, doing all this with strangest and most varied new methods of cruelty, never seen or heard of before."[31] Similarly, the American dogma of Manifest Destiny in the early nineteenth century justified the aggressive territorial expansion of a land-hungry European emigrant populace at the expense of native tribes, by presuming a moral superiority over "the savage Indians."[32] From the Holocaust to the hunting of Bushmen by Boers to Bosnian ethnic cleansing: Further examples need not be elaborated. Often the ingroup-outgroup bias appears brutal to the level of bizarre absurdity, while other times it looks ridiculously trivial, for instance when fights arise between children arbitrarily assigned to groups given differently colored clothes by their teachers. Either way, it appears swiftly and strongly.

As an emotion promoting intragroup solidarity and intergroup hostility, ingroup-outgroup bias is perfectly expected in a species with a long history of intergroup aggression. Stupid and cruel as it often is, this bias may have evolved as part of the winners' strategy. Darwin put it this way: "A tribe including many members who, from possessing in a high degree the spirit of patriotism, fidelity, obedience, courage, and sympathy, were always ready to aid one another, and to sacrifice themselves for the common good, would be victorious over most other tribes, and this would be natural selection."[33] Darwin wrote that passage to show how morality could emerge out of natural selection for solidarity.[34] And, of course, the concept that moral behavior, the "ingroup" half of ingroup-outgroup bias, has roots in evolutionary history is attractive. But from beneath that attractive idea we can also dig out the unattractive one: that morality based on intragroup loyalty worked, in evolutionary history, because it made groups more effectively aggressive.

Ingroup-outgroup bias contributes to many community crimes perpetuated by individuals. Collective excitement worsens the effect. In 1930 in the American South, African-American James Irwin was lynched on suspicion of raping and killing a white girl. "Irwin was tied to a tree with chains," a historian tells us. "Approximately a thousand people were present, including some women and children on the edge of the crowd. Members of the mob cut off his fingers and toes joint by joint, mob leaders carried them off as souvenirs. Next his teeth were pulled out with wire pliers." So it went. James Irwin was castrated, burned alive, and shot.[35] A typical, horrendous mob scene.

It is particularly horrific to realize or remember that the people who make up such lynch mobs, or who go wilding or gangbanging or assault women at professional conventions, are mostly ordinary people. Next day they may regret their "madness" as they return to their ordinary lives experienced with ordinary emotions. Alcohol may have helped loosen their inhibitions, but alcohol is not necessary. The sheer excitement of the moment itself works like a drug — just something that happens to individuals losing themselves in the excitement of crowds. In losing themselves, though, they also commonly lose their reasoning and surrender to unexamined emotions. Journalist Bill Buford describes what it feels like to be in the middle of a British soccer crowd: "They talk about the crack, the buzz, and the fix," he writes. "One lad, a publican, talks about it as though it were a chemical thing or a hormonal spray or some kind of intoxicating gas." For Buford himself, the feeling was much the same: "I am attracted to the moment when consciousness ceases: the moments of survival, of animal intensity, of violence. . . . What was it like for me? An experience of absolute completeness."[36]

Deindividuation is the formal name for the mindless sinking of personal identity into the group of Us. There is no particular reason for it to appear in the temperament of a species that lacks intense intergroup aggression. But when we view humans at full length, as a smart and upright species emerging into the present from a 5-million-year history of selection for effective intergroup aggression, deindividuation makes perfect sense. Deindividuation produces, in the words of sociologist Georg Simmel, a "noble enthusiasm and unlimited readiness to sacrifice."[37] That it also produces irresponsibility and deeply unpleasant behavior is only relevant from the point of view of Them.

One social science textbook characterizes the "wild, impulsive behavior" and "antisocial actions" of a group this way: "It is precisely in such settings that human beings may turn upon their fellow men and women with a savagery and a brutality unmatched by any other living creature on earth."[38] In reality, that savagery is matched by several other living creatures. Human savagery is not unique. It is shared by other party-gang species. And in those species that share our propensity for tearing an enemy to pieces, we will likely discover, in cruder form, the same processes that cruelly increase a group's effectiveness in destroying a villain.

Our ape ancestors have passed to us a legacy, defined by the power of natural selection and written in the molecular chemistry of DNA. For the most part it is a wonderful inheritance, but one small edge contains destructive elements; and now that we have the weapons of mass destruction, that edge promotes the potential of our own demise. People have long known such things intuitively and so have built civilizations with laws and justice, diplomacy and mediation, ideally keeping always a step ahead of the old demonic principles. And we might hope that men will eventually realize that violence doesn't pay.

The problem is that males are demonic at unconscious and irrational levels. The motivation of a male chimpanzee who challenges another's rank is not that he foresees more matings or better food or a longer life. Those rewards explain why sexual selection has favored the desire for power, but the immediate reason he vies for status is simpler, deeper, and less subject to the vagaries of context. It is simply to dominate his peers. Unconscious of the evolutionary rationale that placed this prideful goal in his temperament, he devises strategies to achieve it that can be complex, original, and maybe conscious. In the same way, the motivation of male chimpanzees on a border patrol is not to gain land or win females. The temperamental goal is to intimidate the opposition, to beat them to a pulp, to erode their ability to challenge. Winning has become an end in itself.

It looks the same with men.

10

THE GENTLE APE

ALTHOUGH HE HAD NEVER flown to Djolu before, our pilot explained, the airstrip was marked on his map. "Up to halfway," he said, "it's very familiar. Mission out there. Small village and a little clinic for pregnant women. They have about a hundred patients at a time. After that there's an hour and a half where there's not a lot of huts and villages, just broccoli — that's what I call trees. And then you land at a little strip in the middle of the forest."

The weather report was relayed our way: "The rain has stopped. There's a medium cloud layer with some breaks, and now they can see a higher level." And so we climbed into the plane. The pilot pointed out the airsickness bags, strapped on a white crash helmet, passed out chocolate cookies, turned on both engines, and we bumped along the strip, gathered speed, lifted and rose and turned over Lake Kivu, then headed west above the treetops. We passed over a town, some villages, a few broken roads and some tiny villages, then green forest and a hazy horizon.

After an hour and a half we corkscrewed down and landed at the mission airstrip to make the once-every-two-months mail delivery. Then we took off again, rose and flew west across forest and an occasional river for most of the next two hours. Gradually, we began to come across a few signs of human presence. The pilot's map, spread out across his lap, showed little circles and squares marked "church" and "clearing" and "plantation," and then some village names: Bolingo, Yaleta, Itenge, Bumbo, Djolu. We flew above a few small circles and squares; and the pilot crayoned down on his map all good hints and clues for his own return. He wrote: "hog house" and "metal roof" and "string of houses" and "Big Hut."

At last we located a dirt track and some brown holes in the forest. We dipped beneath clouds, passed into bumpy air, crossed over some farm bush, moth-eaten forest, a thatched house, smoke, tin roofs, rectangular houses, then a whole village. The plane fell and banked into a cushion of air, and suddenly we were below the trees and dropping directly onto a patch of dirt scraped out of the forest. Djolu.

In Djolu we were detained by the district commissioner during eight hours of disputatious negotiation over how large a fee the return of our passports commanded. Permission to cross the four-log bridge out of Djolu was granted that evening, and we drove a borrowed vehicle down the 80-kilometer track to the village of Wamba and Takayoshi Kano's house and research site.

By day the village of Wamba can be found, still and sun-beaten, on either side of the track from Djolu. In grass at the edge of the road, white butterflies flutter and scatter in the breeze. The Kano house, mud brick with a tin roof and a yellow rainbarrel, sits to one side of the road. A chicken picks through kitchen scraps in the doorway. To the other side, three dozen boys and young men of Wamba play serious soccer in an open field. The smell of woodsmoke rises in the air.

This is the world of the Mongandu. Stopped by poor roads from selling crops or getting jobs, these people have returned to old ways, hunting hard. Elephants, leopards, monkeys, antelope, the bigger birds: Virtually all are gone. But it's still a magical forest here. The Mongandu believe that humans and bonobos once lived as brothers,[1] and so they have never allowed hunting of bonobos.[2] You can walk into the forest for hours without seeing any large animal. You wonder if you will find bonobos before the heat and humidity exhaust you. And then at last you hear what at first you think are birds in the tree crowns, a chorus of high voices, of chirps and squeaks and shrill soft screams....

In the fall of 1928 Ernst Schwarz, a German anatomist who knew primates well, spent a few weeks examining new material in the Congo Museum at Tervuren, Belgium. There had been a flurry of recent acquisitions. Among them were chimpanzee skulls that had been sent from the Congo — modern-day Zaïre — in three separate shipments in December 1927. But Schwarz noticed something strange: One of the chimpanzee skulls was curiously small. He looked more carefully. Within a few weeks, on October 13, 1928, Schwarz announced to a meeting of the Cercle Zoologique Congolais the existence of a previously unknown form of chimpanzee, an entirely new species he named *Pan paniscus*, the pygmy chimpanzee, or, as it was subsequently to become known, the bonobo.[3]

So, half a world away from where they live, and almost fifty years after a specimen had first been deposited in a European museum,[4] bonobos[5] quietly entered the world of modern science.

The reason it took so long for bonobos to be discovered is simple. They look very much like chimpanzees. Even for an expert the two species are easily confused. The greatest of America's first generation of primatologists was Robert Yerkes, author of such seminal works as *The Great Apes* and *Chimpanzees*, and founder of what is now the world's most important primate research center.[6] Yerkes loved apes on a personal as well as a scientific level. He raised chimpanzees and bred them, and even made pets of some of them. One of Yerkes's most celebrated apes, Chim, is now agreed to have been a bonobo,[7] but the species was not recognized in 1923 and 1924, when Yerkes spent his time with Chim. Yerkes kept Chim at his New Hampshire farm along with Panzee, a female chimpanzee; both were about two years old, and he marveled at the differences in temperament between them. In a popular book, the primatologist devoted a whole chapter to Chim because "in all my experience as a student of animal behavior I have never met an animal the equal of Prince Chim in approach to physical perfection, alertness, adaptability, and agreeableness of disposition." He noted that all who saw him recognized him as "an unusual type which rarely is seen in America," and he even speculated that Chim's nature reflected "species or varietal peculiarities." Ultimately, of course, Robert Yerkes regarded Chim as a chimpanzee, albeit an extraordinary individual. "Doubtless there are geniuses even among the anthropoid apes," Yerkes concluded, and his beloved Chim was one.[8]

In spite of their original name, "pygmy chimpanzee," bonobos are only a little smaller than chimpanzees. Indeed, the scanty data available indicate that they weigh on average the same as the smallest known chimpanzees, those from Gombe: around 29 kilograms for a female and 40 kilograms for a male. Their heads are small. Their bodies are slender. Their arms and legs are long. The hair on their heads is parted in the center. Their mouths and teeth are small. Their faces are usually blacker than chimpanzees' (who can have pale faces, especially as babies). They have pink lips. These features identify a bonobo.

The fascinating thing about bonobo anatomy is that it enables us to look back at their ancestral relationships with chimpanzees and gorillas. Bonobos, for example, have a long, thin shoulder blade compared to a shorter, broader version in chimpanzees and gorillas. So the unique bonobo shoulder blade evolved only after chimpanzee and bonobo ancestors separated (which occurred long after their common ancestor split from gorillas, of course). The same is true of most other characteristics. It turns out that chimpanzees are more gorilla-like than bonobos in their chromosomes, their growth patterns, their blood groups, their calls, and their physical appearance. In their sexual and social behavior neither chimpanzees nor bonobos are very much like gorillas, but in both cases, as we shall see, bonobos represent an exaggerated form of the chimpanzee system. Earlier we noted that chimpanzees are a conservative species, but we didn't mention bonobos at that time. Now we see one more piece of evidence of how conservative chimpanzees really are, because bonobos as well have changed rapidly away from the chimpanzee mold. To put it another way, bonobos are best imagined as descended from a chimpanzee-like ancestor, rather than chimpanzees from a bonobo-like ancestor. Genetic dating puts the split some time between 1.5 and 3 million years ago.[9]

Though bonobos are clearly distinct from chimpanzees, the physical differences between the two species are less than the average differences between many populations of humans. But walk with them in the forest, and the differences shine clearly. Chimpanzee parties communicate to each other with huge, throaty screams, hoots and barks that can carry for a kilometer. The equivalent calls of bonobos are brief, high-pitched, soft hoots, reaching much less far. To Kano, the first primatologist to hear bonobos in the wild, they sounded like birds: like "hornbills twittering in the distance."[10]

The gentleness of bonobo calls is only the first of an extraordinary suite of behavioral differences that are now known to divide the two species. In the twenty years since the Wamba studies started, Takayoshi Kano, Suehisa Kuroda, and their team have added much data to that already known from captive studies to show how bonobos are unique. This pioneering work has been confirmed and amplified in other sites. By 1992, when Kano published the first book in English describing a field study of bonobos,[11] the story of bonobo life could be told with confidence. It is a tale of vanquished demonism.

As we enter the social world of bonobos, we can think of them as chimpanzees with a threefold path to peace. They have reduced the level of violence in relations between the sexes, in relations among males, and in relations between communities.

First, how do males treat females? The basic evidence is clear and strong. Among bonobos there are no reports of males forcing copulations, battering adult females, or killing infants.

On the surface, bonobos have a social life very much like chimpanzees, living in communities, sharing a range with eighty or more others, traveling in various-size parties within the community range, living with their male kin group, and defending their range against outsider males. Most importantly, bonobos have the same size difference between males and females as chimpanzees. So why don't bonobo males exert their physical power over females in the way chimpanzees do? The answer takes us into the heart of bonobo society.

Among chimpanzees every adult male is dominant to every adult female, and he enjoys his dominance. She must move out of his way, acknowledge him with the appropriate call or gesture, bend to his whim — or risk punishment. The punishment by a bad-tempered male can vary from a hit to a chase through trees and along the ground, until the female is caught and pulled and kicked and hit and dragged, screaming until her throat cramps, reminded to respect him next time.

But among bonobos, the sexes are codominant. The top female and the top male are equal. The bottom female and the bottom male are equal. In between, your rank depends on who you are, not what sex you are. Of course, just as among chimpanzees, social rank is not the only thing that determines if you get your way. At least as important is who will help you.

In 1982 Ude was the second-ranking male in Wamba's E-group; and Aki was one of the most powerful females, close to the top rank. Her son was a young adult, testosterone rising in his veins, starting to challenge the older males. One day Aki's son charged aggressively at Ude, screaming and dragging a branch, swerving away only at the last second. Ude, clearly agitated, slapped his challenger before being calmed by an intervention from the top-ranking male. But Aki's son charged again. This time Ude chased him, but Aki's son stood his ground. A fight ensued, with some kicking and hitting. The tide turned when Aki joined in. Carrying her screaming baby on her belly, she chased Ude not once but a dozen times. Other females joined in the fray with supportive calls. It wasn't long before Ude fled. Ten years after that single incident, Ude is still subordinate to Aki's son: fleeing, presenting, or taking little steps away whenever the two meet.[12]

Bonobo sons are almost inseparable from their mothers. They groom with them more than with males or anyone else.[13] They stay with them throughout their lives, consistently in the same party. From the sons' point of view, this makes sense because their mothers' support appears to be crucial for success in competing with other males. Males whose mothers are alive tend to be high-ranking. Kano's study documents four young males who rose rapidly in rank because they had mothers alive and supporting them; two other males fell in rank when their mothers died. In another case a reversal in dominance between two mothers resulted in a corresponding reversal in the ranks of their sons.

Now it's no surprise that a mother would support her son. It happens in chimpanzees and many other species. But for most species this support does little for the son's status. A mother rallying for her son rarely makes a difference among chimpanzees, for example. And so male chimpanzees rarely follow their mothers; if an association between them is deliberate, it will normally happen because she follows him. But in the case of bonobos, the support of the mother really does matter.

The reason why mothers are so valuable to their sons opens a window into the bonobo world. Female bonobos cooperate with one another in ways males do not. Among bonobos, the mother-son relationship is the closest bond there is between males and females; and if a mother calls for help, other females will respond. So if a son or his mother is harassed, the mother's group of females is always liable to counterattack in her support —

just the sort of thing that happened when Aki defended her son against Ude. The effect of the massed rally is overwhelming. Female power wins. By contrast, the males never cooperate with each other, either to defend themselves or to attack females. Thus, even the highest-ranking male can be defeated when females gang up on him.[14]

So the Wamba data suggest that female power is the secret to male gentleness among bonobos. However, the females don't often have to assert their power. Occasionally a male may lose his temper, attacking a female badly enough to tear her ear, for example. But such cases, as Kano's research makes clear, are very unusual. Males rarely attack females (half as often as they attack each other),[15] and when they do, they are liable to be driven off by a gang of females.

Not just in defense do females use their power. Chimpanzees and bonobos both occasionally find foods that they especially prize. Two types of food in particular are valued: huge fruits, such as the soccer ball–size wild jackfruit, *Treculia africana,* and meat. When chimpanzees or bonobos come across these very valuable foods, one individual normally manages to become the temporary "owner," while others gather round begging for shares. For chimpanzees in a mixed-sex party, it is *invariably* a male who ends up as the owner, regardless of who found the prize first. But among bonobos either sex can be the owner.[16]

Captive studies tell a similar tale. In Germany, a researcher compared chimpanzees and bonobos by secreting delicious foods such as honey and milk in an artificial "fishing site," where the apes had to spend time slowly extracting the prize by dipping straws and licking them. In the chimpanzee group, the male always dominated all females and monopolized the goodies. But in the bonobo group, females formed coalitions, isolated the male, and then rewarded themselves with the sweet liquids.[17] Cooperation among the females kept the male in his place.

All available observations tell the same story. Female bonobos turn the tables on males. And if bonobo males throw their weight around and become overly aggressive, they are liable to be suppressed by females. The big question, therefore, is what bond makes the females such reliable, predictable supporters of each other. It's not kinship. Bonobos are like chimpanzees in that regard: When a female enters adolescence she leaves her family, migrates to a new community, and settles there. Most of the females she will spend her life with are unrelated to her. No, the bonds among females come not from kinship but from experience. In other words, the newly arrived adolescent must work to develop her support network.

The pattern has been witnessed for only a few females, but it is striking.[18] The most confident descriptions come from Gen'ichi Idani's observation of three adolescent females transferring from another community into E-group at Wamba. Upon arriving at E-group, each of the adolescents targeted a particular (and different) adult female. The adolescents started out by sitting close, alert to their target adult, clearly subordinate but showing a quiet interest. The adolescent initiated most interactions, perhaps when the older female signaled willingness for the interaction to happen. Then, over the course of a few weeks, the pair became each other's most frequent partners in friendly interactions.[19] Constantly alert to the glances and intentions of a particular other, shyly waiting for a signal to approach and be friendly. . . . It looks like falling in love.

Of course, we don't know what the adolescent bonobo feels, so if you don't like the idea of her feeling romantic, let's stick to the facts. What do her friendly interactions with the senior female consist of? Partly it's the ordinary social life of primates, sitting close and grooming each other. But in addition, she behaves, well, romantically. She has sex with the older female.

Researchers describe sex between female bonobos with an accurate but to us unsettlingly clinical term: *genito-genital rubbing*. But the term GG-rubbing (as it is usually abbreviated) hardly captures the abandonment and excitement exhibited by two females practicing it. So let's use the Mongandu expression to describe this remarkable act: *hoka-hoka*.

Here's a typical case of *hoka-hoka*. The adolescent female sits watching the older one. When the older female wants *hoka-hoka* and has seen that the adolescent is waiting, she lies on her back and spreads open her thighs. The adolescent quickly approaches and they embrace. Lying face to face, like humans in the missionary position, the two females have quick, excited sex. Their hip movements are fast and side to side, and they bring their most sensitive sexual organs — their clitorises — together. Bonobo clitorises appear large (compared to those of humans or any of the other apes) and are shifted ventrally compared to chimpanzees. Kano believes their location and shape have evolved to allow pleasurable *hoka-hoka* — which typically ends with mutual screams, clutching limbs, muscular contractions, and a tense, still moment. It looks like orgasm.[20]

Through sitting together, grooming, and *hoka-hoka*, the bond between the adolescent and the older resident female deepens. In a matter of a few months, the adolescent has a "friend" — in the technical sense of an individual with whom she has a specially affiliative relationship. With the development of that friendship, her integration into the new community has begun. It will be exciting to learn more about what happens to allow an adolescent to widen her network of support. For the moment, little is known. We do know that all the females in a bonobo community have close affiliative relationships with other females, consistently expressed through *hoka-hoka*. And we know that the dominant females respect their subordinates. Unlike males, they don't display aggressively at each other, and the subordinates rarely give submissive signals (such as grunts or screams). Bonobo females groom each other more than chimpanzee females do, though not as much as females of either species groom males. Aggression occurs among the females, but it is rare.[21] When there is tension, they tend to make up quickly. Relations among females make for a generally peaceful life, in which the seniors do little to exert the authority of their position. It looks as though the adolescent's development of a bond with a senior female is her passport into a network of support and security.[22]

As for how males treat other males of the same community, bonobos and chimpanzees show lots of similarities. In both species, males compete for status and form hierarchies, with the alpha male particularly easy to recognize.[23] In both, males stay close and groom each other often, for similar amounts of time.[24,25] In both, the males try to intimidate each other by charging while dragging branches. And, in both, they attack each other more often than they attack females. They can bite and hit and kick and slap and grab and drag. After aggression, the males of both species make peace, with the antagonists approaching each other and doing something friendly, like grooming.

But there are many differences, too, and they all tell the same story. Whereas chimpanzee males are prepared to fight fiercely and risk a good deal to attain the alpha position, bonobos are not. Male bonobos just don't seem to care quite so much about being the boss. Bonobos fight less often, with lower intensity, and have less elaborate behaviors for preventing or resolving their differences. What researchers call "attacks" among male bonobos are typically displays of various kinds — mostly charges — without physical contact. The unseating of an alpha male from his position of highest dominance can lead to life-threatening wounds among chimpanzees,[26] but no such wounds have been seen among competing bonobos. In other words, bonobo aggression tends to be much less severe.

The contest for dominance is more elaborate among chimpanzees. Chimpanzees have ritualized signals of status recognition; bonobos don't. For example, in chimpanzees, signals of reassurance or reconciliation exaggerate the hierarchy — such as when a subordinate crouches and approaches a dominant while panting softly, and the dominant reaches out and pats the outstretched hand of the subordinate. But all of the equivalent behaviors among bonobos are symmetrical — for example, two males mounting each other in turn, typically with obvious sexual excitement. And aggression among bonobos often leads quickly to resolution, whereas chimpanzees take time to make up.[27]

Chimpanzee males form alliances, as we have seen, which are crucial for their success in gaining and keeping high rank. Bonobo males don't. Along with this distinction come several differences in political strategizing. Even though the amount of grooming is similar for the males of both species, the pattern is different: Chimpanzee males tend to groom all their fellows, whereas bonobo males show clear favorites. Why? One theory is that chimpanzees use grooming as a way to curry favor with rivals rather than simply expressing friendly preferences.[28] Chimpanzee males do the same with food sharing; unlike bonobos, they share as a political tactic.[29]

Milder attacks between males, less male competition for rank, no male alliances to gain political advantage — why don't male bonobos care as much as chimpanzees about being the big cheese? The influence of mothers and the female firepower is obviously very strong. But bonobo males are less aggressive with each other for another reason as well: They are much less concerned about who mates with the females. Among chimpanzees, copulation attempts by low-ranking males are often stopped by high-ranking males, especially near ovulation time. This happens very rarely among bonobos.[30]

Why don't bonobo males care more about who gets to mate? The answer looks simple. Males can't tell when females are ovulating, apparently because the crucial smells that signal the approach of ovulation to male chimpanzees are simply absent in female bonobos.[31] This ignorance is crucial. Ovulation is the time when the egg slips out of the ovary and readies itself to be fertilized. It happens once a month for a female, who thereby becomes ready for impregnation. By the logic of natural selection, a female's moment of ovulation is the critical time for males to mate with her and, of course, to stop other males from mating. Chimpanzees appear able to tell the day of ovulation rather precisely, and males compete intensely for copulations at that time. Among bonobos, males show more interest in females during cycles when ovulation is relatively likely to occur. But during those cycles, no researcher has yet reported increased male anxiety about mating access as the specific day of ovulation approaches. It looks as though females conceal their ovulation.

Humans often proudly consider themselves the sexiest primates. But consider bonobos: They can mate dozens of times a day; males and females eagerly engage in heterosexual and homosexual sex; they manipulate each other's genitals with hands or mouth; they adopt an impressive variety of copulatory positions; their genitalia, male and female, are proportionately larger than humans'; and they begin having sex long before the onset of puberty — from about one year old.[32]

But more interesting perhaps is what bonobos do with their sex. Many people assume that St. Augustine was right that humans are the only animals who challenge "the inseparable connection . . . between the two meanings of the conjugal act: the unitive meaning and the procreative meaning."[33] Alas for St. Augustine. He should have gone to Wamba. Bonobos use sex for much more than making babies. They have sex as a way to make friends. They have sex to calm someone who is tense. They have sex as a way to reconcile after aggression. These three patterns have been beautifully documented in captivity,[34] and observations in the wild suggest that at least the first two, and probably all three, occur there, too. We know their sexual activity has nothing to do directly with reproduction in many cases, if only because many of their sexual encounters involve homosexual pairs or sex with infertile infants. And heterosexual sex between adult bonobos as well often looks as if it is taking place for nonreproductive reasons. In other words, just as people use sex as a way for deepening relationships, comforting each other, and testing each other, not to mention having fun or getting pleasure, so do bonobos.

Such a diverse sexuality among these apes both amazes us and leads us to wonder why it evolved. Again, a comparison with chimpanzees offers some insights. When a female chimpanzee has her monthly sexual swelling, she is subject to being herded and attacked by males. She has to dodge the noisy, dangerous fights of males challenging each other. She looks emotionally stressed, spends little time eating, and suffers wounds. Vulnerable to male power, she can pay a high price for being sexy. But since female bonobos are able to control males, their sexual attractiveness is not a liability but a strength, particularly because with the disguised ovulation time, males no longer know precisely when it's best to compete with each other.[35]

The third part of the threefold bonobo way to peace was the last discovered and remains the least understood. Intercommunity violence is reduced; some encounters between communities are even friendly.[36] True, smaller parties normally avoid larger ones, and if parties from different communities meet, there can be a fight. But when relaxed meetings between communities do happen, how remarkable they are.

These meetings happen at the borders. At Wamba, on December 21, 1986, Gen'ichi Idani was sitting at a clearing where sugar cane had been strewn about to attract members from one of the two communities whose ranges bordered that part of the forest. Idani sat listening for the calls indicating that a party of bonobos was on its way. Unusually, he heard two parties calling loudly from opposite directions, simultaneously approaching the feeding area. A few minutes later, Idani saw the two parties emerge from either side of the clearing and move slowly toward the sugar cane, looking at each other. Idani recognized the individuals of one party as members of E-group; the others were from P-group.[37] Gradually, individuals from the two parties sat down within a few yards of each other, continuing to call, not fighting but not mingling either. It was a standoff, with the two parties separated by a sort of demilitarized zone. And then, after thirty minutes of this strange truce, a P-group female crossed the neutral ground and had *hoka-hoka* with a female from the other community. What followed was unprecedented for ape watchers. For the next two hours the two parties fed and rested together almost as if they were members of a single community, with only the mature males of the two groups still quietly retaining their old social boundaries.

In the next two months, this same scene was replayed, with

variations, some thirty times. E-group and P-group had thirty-two and thirty-nine members, respectively, so there were sometimes as many as seventy individuals gathered into a small area. Some days they even traveled together for a few hours before separating into sleeping parties, then recombined the next morning.

Friendliness was always initiated by the females. One particularly friendly individual was a P-group female who had grown up in E-group. She often approached E-group females and was also approached by them. They seemed to remember each other well. In addition, adolescent females seemed very interested in adult females from the other group, as if they were anticipating a possible transfer into that group at a later time. Females groomed with members of the other community; they continued their intercommunity *hoka-hoka*; and they also copulated with males from the other community. Most remarkably, the males simply watched, inert, while the females of their community copulated with males from the other side. I was sitting with other chimpanzee observers when Idani played videos of these sequences to the International Primatological Society in Japan in 1990. It was hard for any of us to believe that the bonobo males were from separate communities. Nothing I had seen before showed how different bonobos were from chimpanzees.

The Wamba study started in 1974, and it has been continuous since then. For many months, sometimes years at a time, however, political conditions in Zaïre have prevented outsiders from watching directly. But Takayoshi Kano and his team have hired and trained several Mongandu men to observe, such as Norbert Likombe Batwafe and Ikenge Justin Lokati, the two experts who led us in the forest. These men provide the continuity, the long-term evidence: twenty years of information with up to four different bonobo communities watched in a year. They have often seen bonobos run to the border to chase away neighbors. Clashes can result, sometimes leading to bloody wounds, so bonobos don't live in Utopia. But in all that time no one has seen border patrols, raiding, lethal aggression, or battering of strangers. The difference from chimpanzees looks clear.

Three key behavioral comparisons lead to the same conclusion: Male bonobos are not as violent as male chimpanzees.[38] But there is a final and especially intriguing difference between the males of the two species.

Male chimpanzees everywhere hunt and kill mammals. A lone individual may grab an unguarded piglet, or a party of males may collaborate to drive monkeys through the treetops toward an ambusher waiting hidden behind a tree trunk.[39] They can kill often. In one extraordinary two-month period, the main community at Gombe killed seventy-one monkeys in sixty-eight days. For twenty years the Gombe chimpanzees killed so many red colobus monkeys that about 30 percent of the colobus population was eliminated each year.[40] Hunting is a regular feature of chimpanzee life at all the research sites where monkeys are present, Gombe, Taï, Mahale, and Kibale.[41] And in every place, the chimpanzees' visceral reaction to a hunt and a kill is intense excitement. The forest comes alive with the barks and hoots and cries of the apes, and aroused chimpanzees race in from several directions. The monkey may be eaten alive, shrieking as it is torn apart. Dominant males try to seize the prey, leading to fights and charges and screams of rage. For one or two hours or more, the thrilled apes tear apart and devour the monkey. This is blood lust in its rawest form.

Bonobos like meat, too. Like chimpanzees, they are ready to grab and eat small antelope infants.[42] They eat flying squirrels and sometimes earthworms.[43] But remarkably (since they are so similar to chimpanzees in their body size and physical abilities), they have never been seen eating monkeys. They should surely be able to catch monkeys — in fact, we know they can, because it's been seen three times at Lilungu, another Zaïrean study site.[44] Once an adult male gently flailed a juvenile Angolan colobus monkey by its tail, then pushed it roughly onto a branch before allowing the primate to leap away, apparently unharmed.

A week later the same male held and carried a young redtail monkey for two hours. For the first twenty minutes the monkey was alive, but while the bonobos were out of sight of the observers, the monkey died, probably from rough handling. A few weeks later, another male bonobo, Lokwa, carried a young redtail again. The redtail was patently scared when first seen in Lokwa's hands. It shrieked and tried to escape, grabbing at passing branches until Lokwa struck the monkey and forced it to let go. Lokwa was fascinated by this strange new plaything. He held the monkey down, bent and stretched out its legs, and groomed it. He gripped the monkey's hand in his mouth, exploring and nibbling at it. He thrust at it with an erect penis. He turned somersaults while hanging on to the monkey's tail. He leaped about carrying it, careless of the blows his monkey received to its head. He climbed with his monkey pressed to his chest, like a mother carrying her baby. He did all these things for just over an hour before disappearing into the thick forest. An hour later, when the observers had found Lokwa again, the monkey was dead — but Lokwa continued to play with the body and carry it with him for the rest of the afternoon, and for at least three hours the next morning. In these three cases, the bonobos tried to get the monkeys to play. The observers thought the bonobos were treating their monkeys like dolls or pets, not prey.

Red colobus monkeys provide the most interesting comparison here. Red colobus are the commonest prey of chimpanzees at the four sites where predatory patterns are best known (Gombe, Mahale, Kibale, Taï), as well as the commonest prey monkey of human hunters in lowland African forests.[45] It doesn't surprise me that chimpanzees show such a lust for red colobus meat. I have eaten no mammal meat since 1977, but to find out how red colobus tastes I chewed recently on the meat of a carcass dropped by Kibale chimpanzees on a day when they were satiated from multiple kills. To the vegetarian me, the taste was too meaty, but to the lingering meat-eater in me, the scrap was tender, sweet, and delicious. How remarkable, then, what scientists have seen when bonobos encounter red colobus. Only three interactions are known. All three occurred at Wamba in the same month, and all were friendly. Four adult male colobus monkeys were involved, solitaries or members of an all-male troop. They approached bonobos, and twice groomed young apes. No aggression was seen.[46]

Surely this means that red colobus are not afraid of Wamba bonobos and therefore must not have been hunted by them. If the colobus had been hunted, we would expect the relationship to be far more aggressive on both sides. Red colobus that live in the same areas as predatory chimpanzees defend themselves well. In Kibale they sometimes even attack chimpanzees without provocation.[47] The lack of any overt aggression between red colobus and bonobos must mean that red colobus are not vulnerable to predation by Wamba bonobos. Red colobus are rare at Wamba, but even so, it is remarkable that they should be so unafraid.

So bonobos like to eat meat, can catch monkeys, and live in areas that include chimpanzees' favorite prey. Yet we find no indication — at Wamba or Lomako[48] or Lilungu, the three sites where something is known about meat eating among these apes — that bonobos ever prey on monkeys. Most odd. Since bonobos have descended from a chimpanzee-like ancestor, they surely hunted monkeys in the past.

As to why they don't hunt monkeys now, some relatively simple or uninteresting explanation is possible. For example, maybe the particular populations of bonobos studied so far are exceptional, and by sheer coincidence none has developed a habit or tradition of preying on monkeys. In a similar way, the chimpanzees in West Africa's Taï Forest collaborate so effectively in hunting that they kill monkeys as often as anywhere known, but nevertheless do not kill antelope. Instead, they play with them, much as Lokwa played with the redtail monkey. If there is cultural variation of this sort — rather than an evolved and inheritable pattern — other bonobo populations will be found that buck the trend and do prey on monkeys. But perhaps something deeper is involved.

In 1953 Raymond Dart, the discoverer of *Australopithecus*, argued that carnivory in human ancestors generated the "loathsome cruelty of mankind to man."[49] In this now unfashionable view, Dart thought that the taste for animal meat led inexorably to cannibalism, and thence to unspeakable cruelties. The anthropologist waxed poetic about his theory: "The blood-bespattered, slaughter-gutted archives of human history from the earliest Egyptian and Sumerian records to the most recent atrocities of the Second World War accord with early universal cannibalism, with animal and human sacrificial practices or their substitutes in formalized religions and with the world-wide scalping, head-hunting, body-mutilating and necrophilic practices of mankind in proclaiming this common bloodlust differentiator, this predaceous habit, this mark of Cain that separates man dietetically from his anthropoidal relatives and allies him rather with the deadliest of Carnivora." Hot stuff! But Dart's thesis was cooled considerably when Konrad Lorenz and others pointed out that in cats and birds, predation and aggression are directed by entirely different neurological systems.[50] So Dart's heady connections between feeding behavior and personal violence appeared no more than imagination.

And yet, do bonobos tell us that the suppression of personal violence carried with it the suppression of predatory aggression? The strongest hypothesis at the moment is that bonobos came from a chimpanzee-like ancestor that hunted monkeys and hunted one another. As they evolved into bonobos, males lost their demonism, becoming less aggressive to each other. In so doing, perhaps they lost their lust for hunting monkeys, too. It could be that they are less readily excited than chimpanzees by blood, by the prospect of a kill. Or perhaps they are more sympathetic to a victim. Or possibly male coalitionary skills have been lost. The next wave of studies from Zaïre will help us decide if, among the apes, the same cognitive architecture unites predation and social demonism. Murder and hunting may be more closely tied together than we are used to thinking.

11

MESSAGE FROM THE SOUTHERN FORESTS

UNAWARE EVEN OF the existence of bonobos, Robert Yerkes summarized Chim's temperament this way. "Seldom daunted, he treated the mysteries of life as philosophically as any man." This ape, Yerkes continued, "was even-tempered and good-natured, always ready for a romp; he seldom resented by word or deed unintentional rough handling or mishap. Never was he known to exhibit jealousy. If I were to tell of his altruistic and obviously sympathetic behavior toward Panzee I should be suspected of idealizing an ape."[1]

Those who work with bonobos are often suspected of idealizing them. But it's now clear that the bonobos' reputation is fair. Their remarkable qualities of sympathy and restraint lie not just in particular individuals, like Chim, who happen to have delightful personalities, but in the species as a whole. We're dealing with a species that experienced several odd things as it diverged from its chimpanzee-like ancestors, of which the most unusual was the reduction in personal violence.

So naturally we want to know why that happened. What set bonobos off on their curious path after they left their common ancestry with chimpanzees? This chapter looks for an answer embedded in the ultimate needs of daily life. Our search will take us into the African forests and out again on a journey through space and time to look back 2.5 million years. And we will suggest a specific prehistorical link between the evolution of bonobos and the evolution of humans, making us ponder the irony of an event that could have changed two different apes in two very different ways.

We have seen that the power of female alliances explains why bonobo males so rarely exert brute force over females. It also explains why females have been able to evolve their hyper-sexuality. And the reduction in male-against-male violence flows from the males' inability to monopolize females, perhaps also from their ignorance of the females' fertile periods. Additionally, the power of females explains the importance of mothers to adult males. It probably even accounts for why sexual behavior is so relaxed that it has become a medium of communication, not just a way to achieve conception. Female power is a sine qua non of bonobo life, the magic key to their world. So where did bonobos find this key to a better life?

That's one problem. The other has to do with lethal raiding. Even female power can't explain why intercommunity violence between groups of males has been so repressed. Male bonobos still live with their fraternal kin groups in communities that hold land, and they still fight over it with battle cries and even physical engagements. But they don't go on border patrols. They don't stalk into neighboring territory. And, so far as we know, they don't kill each other.

The two problems have a single answer. Party size.

The contrast with chimpanzee parties was evident from the earliest days of bonobo fieldwork. Early on Kano reported that bonobos differ by having larger, more stable parties. His student Suehisa Kuroda concurred. Kuroda found that even when fruit was scarce, bonobo parties stayed large. Party size averaged 16.9

individuals over the year,[2] compared to a paltry and widely fluc-
tuating two to nine individuals for chimpanzees (in six different sites).[3] In bonobo sites other than Wamba, later researchers
found parties to be smaller, averaging around the same size as
chimpanzee parties, and yet their stability and composition remained different from chimpanzees'. Our most secure generalization is that bonobo parties include more females than chimpanzee parties do, and they vary less in size over the year.
Chimpanzee parties are sometimes small, with individuals or
mother-child groups often traveling alone. Bonobo parties are
never so atomistic.[4] To observers in Lomako, bonobos combined
"the cohesiveness of gorilla society with the flexibility of chimpanzee social organization."[5]

Cohesive. Now this is peculiar. Cohesive parties, according
to theory, depend on a low cost-of-grouping. Why should grouping cost less for bonobos than for chimpanzees? Here are two
species with anatomy and body size so similar that bonobo
specimens were tucked away in European museums for almost
fifty years before someone recognized them as distinct from
chimpanzees. The two apes live in forests separated merely by a
river, the Zaïre. Chimpanzees live north of the river, on the right
bank. Bonobos live south, on the left bank. But though bonobos
live on the southern side of the river, they are actually no farther
south than many chimpanzees — both species have equal use of
the equatorial forests where ape foods grow best. Both straddle
the equator because the Zaïre River forms so great an arc around
the bonobo rainforests that bonobos on the equator have chimpanzees living both to the east and to the west of them. As far as
anyone knows, these forests have the same species of trees on
both sides of the river, the same forest structure, the same ape
foods. Can it really be true that this forest island in the heart of
Zaïre, an island the size of California, is so different from the
forests all around it that there and only there the apes find it
easier (or ecologically less expensive) to live in larger or more
stable groups?

To rainforest apes, forests are first and foremost reservoirs of food. As for the favored foods of bonobos, in 1979 Takayoshi Kano summarized his survey of four sites. Bonobo diets, he said, are different from those of chimpanzees because they include "both the [fruit] foods of [chimpanzees] and [the] fibrous foods of gorillas."[6] In the fifteen years subsequent to Kano's original observations, those "fibrous foods of gorillas" have turned up in significant quantities everywhere bonobos have been studied. Bonobos indeed eat more gorilla-style foods than chimpanzees do.[7]

Their gorilla-style foods are the young leaves and stems of herbs on the forest floor. Bonobos snack on them while traveling between fruit trees. These herbs are quite common, a relic of past disturbance, growing in old tree-fall sites or where the forest has been coming back from being felled decades before. For much of their travel, bonobos may walk as a group, though widely spread, with individuals stopping at their own little herb patch and eating for a few minutes before walking another hundred meters to the front of the party and finding another snack. It is the same way gorillas feed. And just as with gorillas, those herbs are so common that the presence of others in the group does little to reduce the food supply.

Because they have this snack food, the costs of traveling with other individuals are reduced. Bonobos don't have to visit many extra fruit trees every day to satisfy their hunger. The gorilla foods buffer the effects of seasonal fruit shortages and allow bonobos to travel with their fellows more easily than chimpanzees can afford to.

So the diets of bonobos and chimpanzees are different in an important way. The beauty of this difference is that we can explain why it's there. The puzzle, remember, was that the forests of bonobos on one side of the Zaïre River and those of chimpanzees on the other side look so similar. Well, floristically, they are indeed similar.[8] But the important difference between the forests of bonobos and the forests of chimpanzees lies not in the plants; it lies in the animals. The lowland forests occupied by chimpanzees are shared with gorillas. But bonobos have no gorillas in their range. In other words, gorilla foods are commoner in the bonobo forests simply because there are no gorillas to eat them.[9]

You might wonder whether, in the range of chimpanzees, gorillas are sufficiently abundant to make a difference in the food supply. They are, absolutely. Biomass density measures the total weight of a species in a given area, so it provides a way to estimate any species' impact on its food supply. One set of comparisons calculated the biomass density for gorillas to be around 35 to 55 kilograms per square kilometer, compared to 30 to 80 kilograms for chimpanzees in the same forest.[10] For every chimpanzee eating chimpanzee foods in the forest, therefore, there was roughly an equivalent weight of gorillas eating gorilla food. Remove those gorillas from the forest, and chimpanzees — if they were inclined to seek it out — would suddenly discover a lot of gorilla food.

We are close to the end of this winding road. Bonobos can afford to live in larger, more stable parties than chimpanzees because they live in a world without gorillas. They have evolved to take advantage of the more digestible parts of the gorilla diet — not the tough, low-quality stems that occur in patches around swamps, but the juicy, protein-rich growth buds and stem bases of young herbs. We even can see the marks of this evolution in their teeth: Bonobo teeth have longer shearing edges than those of chimpanzees, adapted for eating herbs in a way that surprised people when they first discovered it in 1984.[11] Bonobos have evolved in a forest that is kindlier in its food supply, and that allows them to be kindly, too.

But why are there are no gorillas in the land of bonobos? Why do chimpanzees and gorillas both live on the right bank of the Zaïre River, while only bonobos live on the left bank?

None of the living African apes has a recent fossil record — tropical forests are hostile to the preservation of skeletal remains — so we are forced to speculate. But some basic facts can help focus our speculation. Chimpanzees and gorillas have that old common ancestry, having separated from each other 8 to 10 million years ago. Bonobos and chimpanzees separated from their common ancestor around 1.5 to 3 million years ago. The Zaïre River appears by comparison very old: The rocks in its eastern wall are around 3 billion years old. The forests around it have sometimes been vast and continuous, allowing apes to migrate all the way round. Indeed, many species of mammals live on both sides of the river.[12]

These basic facts mean that some scenarios are unlikely. For instance, it's unreasonable to think that the first apes to reach the river's southern bank were bonobos. Gorilla and chimpanzee ancestors surely lived north *and* south of the river for much of the last 8 million years. Let's not think that bonobos evolved somewhere else and then arrived in the forests to find a world magically without gorillas or chimps. Instead, we must imagine that bonobos evolved *within* the southern forests out of their original chimpanzee-like ancestor — and they evolved there once the ancestral gorillas had left.

To think about why ancestral gorillas left this area, we must step back for a moment and consider climate. Africa's climatic history during the last few million years tells a story of irregular drought. During the ice ages, when so much of the earth's moisture was locked up in the great ice caps, forests dried up and contracted to tiny areas in East and West Africa, surviving only where it was wet — usually where there were mountains, or in river gullies. It is easy to imagine that in a cold, dry period around 2.5 million years ago, African forests were so reduced that gorillas, crucially dependent on the herbs of the moist forest, were forced to retreat along with their forests and finally could survive only in forested mountains, as some gorillas do today. On the Zaïre River's right bank, in fact, forests would have receded west and east into mountains.[13] But on the left bank, there was a problem. No mountains. Here, the low altitude of the Zaïre basin spelled a temporary end to gorilla foods and the permanent end of the southern gorillas. So ancient climatic events probably solve the modern puzzle of how gorillas are distributed in Africa, separated by a 1,000-kilometer gap in the rich heart of the continent.[14]

Chimpanzees today can live in dry areas, uninhabited by gorillas, by eating fruits in the strips of riverine forest that linger amid open savanna. Ancestral chimpanzees could have survived in the same way on the Zaïre's southern side during some appalling 10,000-year drought 2.5 million years ago. Then, when the drought ended, they would have found themselves in a new world. The southern forests would have returned, botanically much the same as before, but without gorillas. Now the southern chimpanzees could exploit the plentiful herbs that gorillas had monopolized as a food source. Herbs became this ape's reserve food. And, with a new and predictable reserve food in place, they could expand their ecological niche, endure well during the fruit-poor seasons, and travel in more stable parties. Stable parties meant they could become bonobos.

Party stability, in other words, produced female power. For females to develop supportive relationships, they need to spend time together.

Even among chimpanzees, there are females who spend time together. Stable female groups are known in the wild, in the strange site in Guinea (West Africa) called Bossou.[15] Here, chimpanzee females do develop mutual support networks, assisting each other in competition for status against other females.[16] In other sites, too, female coalitions spring up occasionally. Pairs of females may be jointly aggressive toward infants, such as the dreaded Passion and her daughter Pom who killed several babies in Gombe.[17] And females may defend each other against males.[18] But many chimpanzee females, such as those in Gombe and Kibale, spend most of their time alone. Even when they do spend time in parties they seldom groom or support each other. It's as if they don't spend enough time together to develop trust.[19] It is mainly in captivity, where a chimpanzee group stays together for a long time, that stable female relationships emerge. Given the chance, those captive female chimpanzees make the best of a bad system. They protect each other from the excesses of male violence by developing mutually supportive coalitions against males.[20]

And that same behavior occurs, as we have seen, among bonobos in the wild. Females form the core of bonobo parties, so that in small parties there are more females than males.[21] Females spend their time closer to each other than males do.[22] Females are more likely to be in the center of the party, with males on the outside.[23] And they form alliances that effectively protect them against male aggression.

As for intercommunity aggression, once again, stable party size is essential — but not because it allows particular females to live together. Instead, it affects the power imbalance that most predicts intercommunity violence in party-gang species. We have argued that the crucial feature favoring lethal raiding by chimpanzees is that a party of several males can find a lone individual and thereby attack at minimal cost to themselves. Among chimpanzees an individual travels alone when forced to do so by the exigencies of a poor food supply. Among bonobos, the existence of herbs gives all individuals the luxury of company, so every party remains capable of putting up a good defense; and raiders into neighboring ground will not find a vulnerable loner.[24]

Around the same time that bonobos and chimpanzee ancestors began their evolutionary divergence, 2 to 2.5 million years ago, another great event was unfolding a few hundred kilometers away, in the savanna woodlands. One line of woodland apes was evolving into humans. Most likely, a drying event led to the loss of fruit trees and thereby pressed one particular population into a full commitment to terrestrial life.[25] This was the start of the genus *Homo*.

Climatic change accelerating, forcing evolution — the ultimate root of many new species lies buried in the history of their habitats. Changes in habitat split populations apart, draw them together, and impose new selection pressures. One kind of habitat change eliminated gorillas and thereby made bonobos. Another eliminated fruit trees and thereby made humans. And it is a fair possibility that the same deep drought, squeezing the African ape with enough pressure to produce two new forms, created humans in the savannas and bonobos in the forests.

The contractions and expansions of the African forest resulted from long-term fluctuations in the earth's climate, cooler during periods when the earth moved farther from the sun, warmer when it was closer. And those changes in yearly average distance from the sun mark changes in the planet's orbit around the sun, as it rocks complexly from symmetrical to elliptical in tune to the pull of other planets.[26] Those distant bodies, in short, steered the climatic change that brought extinctions and new species; and so the origin of both humans and bonobos may have begun partly in the same silent passage of the same flying planets.

Even if these two evolutionary events were forced by different climatic sequences, or if the drying events were less important than we believe, the general principle of the parallel evolution of bonobos and humans remains the same. On different sides of an ancient stream, two different aspects of the ancestral ape emerged.

In the savanna woodlands our ancestors acquired the simple physical hallmarks of being human. Those hallmarks aren't what we normally think of when we talk proudly of "our humanity," but they distinguish humans in the fossil record. The brain expanded beyond the ape range; teeth shrank from the giant molars of the woodland apes; and the skeleton became committed to terrestrial locomotion. Perhaps then our ancestors stopped feeding in trees, but our line kept such ancient climbing adaptations as the mobile shoulder joint. Fossils show these physical characteristics had all appeared by around 2 mil-

lion years ago. With their arrival we can look back and say: Then we began our departure from the apes. They mark the start of our physical humanity.

But what about the moral aspects of humanity? Under the stars of the Pliocene nights, sated by a good day's eating and relaxing in the company of playful children, our ancestors would sometimes have reflected kindly on their peers. Many a friendship would have bloomed. But we know of no foods in the savanna woodlands that would have allowed the woodland apes, or their human descendants, to forage in stable parties as bonobos do. A diet of meat and fruits and nuts and honey and roots would demand constant splitting up to find the best patches. If so, females could not effectively restrain the males, imbalances of power would have continued among party-gangs, and heartless aggression would have been employed occasionally, as it is with chimpanzees and modern humans.

The southern forests provide the message that it didn't have to be this way, that there is room on the earth for a species biologically committed to the moral aspects of what, ironically, we like to call "humanity": respect for others, personal restraint, and turning aside from violence as a solution to conflicting interests. The appearance of these traits in bonobos hints at what might have been among *Homo sapiens*, if evolutionary history had been just slightly different.

12

TAMING THE DEMON

Patriotism is love and defense of one's own country, ordinarily considered a high virtue. The national flag is a sacred symbol; hearts quicken at the sound of martial bands. Celebrated in language, music, and the visual arts, patriotism leads us to some of our greatest acts of heroism. It gives us national holidays and justifies the purest kind of sacrifice. Patriotism can look fine and glamorous — at least in the abstract.

Stripped to its ape essence, patriotism is male defense of the community, gloried among humans and surely enjoyed among chimpanzees and bonobos. For all the cultural determinists' efforts to persuade us that it's an arbitrary choice, patriotism seems such a fundamental aspect of being human that one can hardly imagine how things might be different. And the concept that fighting males are a natural, inevitable part of life is merely reinforced by comparing our own strange mixture of compassion and cruelty with that of chimpanzees. Even the comparison with bonobos emphasizes the idea that male coalitionary violence is primal. After all, the very triumph of bonobo females, in achieving equality with males, is a response to the problems imposed by male coalitions, male violence, and male kinship.

Coalitionary male action in defense of group identity, it might seem, is just an essential building block of social evolution and something that any species must live with.

Yet as soon as we look beyond our little threesome, humans and chimpanzees and bonobos, we quickly discover how odd that system really is. If spotted hyenas could understand the notion of patriotism, for instance, they might laugh uproariously at it. Male defense of the group? What a nonsensical idea when the prime warriors in the clashes of rival clans are females, united through generations of mothers, not of fathers. Any self-respecting hyena clan with some language and a little history would invent female gods, revere female ancestors, and consecrate female principles of power. But male interests? Not on your life!

Spotted hyenas provide a beautiful counterpoint to the chimpanzee-human system, but even among the primates patriotism is unusual. Defense of the motherland, or *matriotism* as we might call it, remains the essential principle among many primates — such as rhesus macaques or savanna baboons — whose females grow, give birth, and die in the troops where they are born. On the rare occasions when these monkeys fight against neighboring troops, the action is dominated by females, fighting in tight coordination with their female kin to defend their troop's status or foraging rights, while the males are only temporary fighters, soldiers of fortune unwilling to risk too much, always capable of shifting allegiance to another troop. Only females, in these matriotic species, are committed by birth and temperament to the fortunes of their troop.[1]

Such comparisons make humans appear as members of a funny little group that chose a strange little path. As we have seen, the evolutionary logic of our odd cluster of ape social systems is explicable, albeit still imperfectly. Ecological pressures kept females from forming effective alliances. With females unable to rely on each other, they became vulnerable to males interested in guarding them. Males seized the opening, collaborated with each other to possess and defend females, and started down the road to patriarchy. We are fascinated by the lives of patriarchal men and the histories of their patriotic alliances — not because we are humans or because we are primates, but because we are apes, and in particular because we are part of a group within the apes where the males hold sway by combining into powerful, unpredictable, status-driven and manipulative coalitions, operating in persistent rivalry with other such coalitions.

Unfortunately, there appears something special about foreign policy in the hands of males. Among humans and chimpanzees, at least, male coalitionary groups often go beyond defense (typical of monkey matriarchies) to include unprovoked aggression, which suggests that our own intercommunity conflicts might be less terrible if they were conducted on behalf of women's rather than men's interests. Primate communities organized around male interests naturally tend to follow male strategies and, thanks to sexual selection, tend to seek power with an almost unbounded enthusiasm. In a nutshell: Patriotism breeds aggression.

Males have evolved to possess strong appetites for power because with extraordinary power males can achieve extraordinary reproduction. Admittedly, not all powerful males have done so. Alexander the Great, arguably the most powerful man the world has ever seen, never showed more than a passing interest in women and fathered just a single child by the time he died at the age of thirty-two. But Alexander bucked the trend. Harems of at least several hundred women have been the norm for the emperors of all major civilizations: Aztec, Babylonian, Chinese, Egyptian, Incan, Indian, and Roman. The women were always young (that is, comparatively fertile), and they were carefully guarded in well-fortified sites, normally by eunuchs. Lest one think these harems existed for mere ritual, or as a random sort of conspicuous consumption without reproductive significance, consider how the harem was organized in the T'ang Imperial court during the golden age of Chinese civilization from 618 to 907 A.D. So many women lived in the emperor's seraglio that meticulous bookkeeping was needed to keep track of menstrual cycles, likely times of fertility, matings, and pregnancies. By the early eighth century the date of a concubine's coupling with the emperor would be stamped on her arm in indelible ink: her ticket to an imperial inheritance, should she give birth nine months later.[2] It's probably fair to say that men with absolute power have routinely fathered several hundred children — rather in the manner of a successful bull elephant seal.

In societies without an absolute ruler, the reproductive benefits of power have been shared between men. War captives, who tended to be women, were routinely given as rewards to warriors and a supporting elite. In the Roman Empire, 10 to 20 percent of the population were slaves, most of them captured during war or born into slavery. And the senior Roman men, though they formally maintained a system of monogamy and legitimacy in order to control the inheritance of their estates and political power, simultaneously practiced a common and widespread polygyny with their female slaves as concubines. Slave women in Rome did little work, bred well, and apparently gave the most powerful Roman men millions of illegitimate children.[3] Again, such use of slaves was the routine for empires.[4]

Any male ruler not trapped by institutional rules has tended to have more wives than ordinary men. This is the logic of sexual selection played out exactly as among many other species: If a male wins power, he will tend to use it to mate as many females as possible. Of course, expanded reproductive opportunity is by no means the only reason why men like power so much. The goal of winning women may be unconscious; or it may in fact have no direct association at all with the emotional systems that drive men to win battles and palace intrigues. Thanks to the effects of sexual selection, men are inclined to seek power for its own sake, even in circumstances where tradition inhibits and law prohibits extra wives or concubines.

Of course, women have also used political power to spread their genes. From the earliest civilizations down to recent times, the historical record shows that powerful women have generally increased their own birthrates by employing wet nurses, thereby reducing the contraceptive effect of nursing. And like powerful men, powerful women have invested heavily in their offspring in order to increase survival rates and the likelihood of their becoming powerful or attractive adults. But nothing a woman can do will expand her reproduction rate in a way equivalent to a man mating with several women.

Because of the large potential reproductive rewards at stake for males, sexual selection has apparently favored male temperaments that revel in high-risk/high-gain ventures. At the individual level, this temperamental quality can show relatively trivial effects. Men may sometimes drive their cars faster or gamble more intensely or perhaps play sports more recklessly than women. But the sort of relatively discountable wildness that, for example, hikes automobile insurance rates for adolescent boys and young men also produces a greater willingness to risk their own and others' lives; and that sort of risk attraction becomes very significant once men acquire weapons. And where men combine into groups — gangs or villages or tribes or nations — this driving, adventurous ethic turns quickly aggressive and lethally serious. Based on this logic, we conclude that imperialism derives partly from the fact that human foreign policy is based on male rather than female reproductive interests.

The idea could be tested by comparing aggressive tendencies in party-gang species according to whether foreign policy benefits females or males. Spotted hyenas' foreign policy benefits females; in chimpanzees it benefits males. The few data currently available support the notion that raiding, an imperialist tendency, occurs more regularly among chimpanzees than among hyenas. But whether or not matriots are less greedy than patriots, the chimpanzee-human system looks clear. The downtrodden of the earth can rail against the imperialism of the temporarily dominant, but imperialist expansionism is nevertheless a broad and persistent tendency of our demonic male species.

What hope, then, for taming the demon?

ENVISIONING FEMALE POWER. In her 1915 utopian novel, *Herland*, Charlotte Perkins Gilman tackled the problem of the demonic male by considering what *Homo sapiens* could become in an imaginary world miraculously freed from the constraints of the male temperament and male-dominated political systems. As the story opens, three American men exploring a remote river somewhere in the wild tropics cross a formidable mountain barrier and stumble upon a society of women only — about 3 million of them living in a country the size of Holland — that has been completely isolated from the rest of the world for 2,000 years.[5]

Herland began as a European-style society resembling ancient Greece or Rome, and organized through patriarchal laws, habits, and institutions, with patriotic kings, generals, and warriors battling each other and dividing up sexual access to the women through marriage, polygyny, and slavery. Geographically isolated on three sides by mountains but still with a narrow pass to the sea allowing trade and communication with the larger world, Herland's historical predecessor was undoubtedly moving in the general direction of the rest of Western civilization, toward something like the unhappy and imperfect present-day Europe and its predictable germination across the ocean, America. Herland, in short, began as a "Hisland" and would probably have developed into something quite like the land the three explorers hail from. Except for a single geologic-historical event.

One day 2,000 years earlier, when nearly all the nation's men were girded for battle and stroking their weapons down at the pass to the sea, getting ready to protect their country from an invasion of similar men from a similar society, a volcanic eruption and earthquake picked up and dropped most of a mountain onto them, burying and killing them at once. The women soon found themselves entirely isolated from any other men by a full circumference of mountains. The discovery of a miraculous procedure for stimulating virgin birth enabled this single-sex society to perpetuate itself over time, while the absence of that other sex proved remarkably liberating.

With the loss of men came a loss of fear. The women of Herland found that with no violent, oppressive, and controlling men around (and the coincidental absence of all wild beasts), they suddenly had nothing at all to fear. Without dangerous men or a need for male protection, the women of Herland no longer had any reason to play men's games: to pretend, for example, that they were weak or incompetent when they were strong and able. And without the constrictions of a patriarchal culture, the women found at last that they could become their true selves, no longer females so much as feminine people, suddenly free to reach their fullest potential according to essential feminine principles now undistorted by the usual masculine demands and expectations.[6]

The essential feminine principle of this utopia is Motherhood with a capital M. The society is fully constructed around child bearing and child rearing as acts of the highest virtue. Of course, motherhood in Herland has little to do with the male explorers' ideas of what mothers ought to do or be. It does not mean that Herlanders belong in the home or that their lives are limited, for example. Without rigidly specifying the details of how maternal love and nurturance are expressed, the Herland program simply elevates them as essential feminine values.

An essential male principle is not spelled out quite so clearly. The three men in this book embody three different responses to the challenge of Herland, but all have come from a society that for more than 2,000 years was organized primarily to serve male needs. In the past it was a violent and starkly hierarchical society run by warlike and dangerous males obsessed by power and their need to control. And we quickly learn that Hisland present — post-Victorian America — is much like Hisland past, a fact confirmed by the three explorers themselves, arriving as they do anxiously gripping their loaded guns, armed intruders clumsily trying to grab some of the first women they meet.

They are quickly taken care of. A highly organized, unarmed but unafraid Herland police force takes them captive by surrounding and overwhelming them. The explorers are then treated like houseguests, and given a year to learn all they can about Herland. What they learn changes them . . . to a degree. But in the end — the last chapter, almost the last page — all three explorers huddle and plot the possibility of returning to this utopia in style: opening up the land to outside commerce, penetrating the great forests at the peripheries, and "civilizing — or exterminating — the dangerous savages," that is, the unspecified native peoples living just beyond Herland's protective mountain ring.

So the problem of male violence is not fully addressed here. Like Gauguin and Melville and Mead, Charlotte Perkins Gilman eliminated male violence from her portrait of an ideal society simply by eliminating males; and we cannot perfectly paste the story's lesson onto an ordinary two-sex society. Nonetheless, the imaginary human society of Herland as well as the actual nonhuman society of bonobos both provide useful tools for thinking. In the case of bonobos, the balance has been tilted from the male-dominated system of the chimpanzees to a male-female sharing of power enforced through female alliances. In the human world, we have no complete, working example of what female power could mean for a society.[7] And thus the importance of the evocative fantasy of Herland is that it, like the bonobo world, allows us to imagine what might happen if women shared power equally with men. Both Herlanders and bonobos guide us into speculating about the meaning of female power, and both suggest the important idea that true female power is not simply a direct or inverted image of male power, but something different entirely, in scope and quality. Among bonobos and in Charlotte Perkins Gilman's imagination, females have power and they sometimes use it in defense against violence. They are matriots, certainly; but they have not become imperialists.

UNTYING THE STRANDS. Animal breeders have readily produced aggressive and pacific strains of various mammals, including mice, rats, and dogs. Although we don't know exactly how temperaments are genetically defined, and although each individual's experience also affects his or her temperament, we have every reason to think that genes significantly influence the development and maintenance of aggression in all mammals, partly because aggressive behavior is increased by the action of testosterone on the brain.[8] And just as nonaggressive strains of other mammals can be bred by artificial selection, so a peaceful strain of human could be bred, too. With some concerted worldwide action we could probably get measurable results within a few generations. Society could, through its own reproductive choices, actually breed a kinder, gentler man — with a temperament less like a chimpanzee's, more like a bonobo's.

Well, obviously it wouldn't work. Persuading the more violent men to abandon hopes of fatherhood would doubtless keep prison builders happy and, in the end, probably engender revolution. But even if the most aggressive, potentially violent men could be persuaded to step aside for the sake of future generations, what about the women? Women's evolved strategic responses to male demonism have included countermeasures and defiance, but they have also included collaboration. That is to say, while men have evolved to be demonic males, it seems likely that women have evolved to prefer demonic males (or imitation demonic males) as mates. This inclination makes good sense in evolutionary terms for two reasons. First, the demonic male is the one who tends to protect the female best from violence by other males and thus keeps her and her offspring safe. And second, as long as demonic males are the most successful reproducers, any female who mates with them is provided with sons who themselves will likely be good reproducers.

In real-life terms this means almost nothing about individual women and their individual choices. It does not necessarily mean that women in general prefer the most violent males. It certainly does not mean that women in general like criminals. It decidedly does not mean that women *choose* to be the victims of male violence. It merely makes sense of the fact that women are inclined to some degree to find attractive some of the marks of male demonism. The marks could be as unsubtle and artificial as the exaggerated body armor pumped up by an ambitious Arnold Schwarzenegger, whose comic book caricature of the demonic male physique and persona appeals to the fantasies of both men and women. Or the marks could be as disturbing and ultimately dangerous as the blustering brutality of Rhett Butler in Margaret Mitchell's *Gone with the Wind*.

In the climactic scene of that book, Rhett — drunk, angry, intimidating, jealous over Scarlett's lingering attachment to Ashley Wilkes — threatens to crush Scarlett's skull "like a walnut."[9] Then, a moment later, he picks her up, carries her roughly up the stairs, as she, "wild with fear," cries out. At the landing, Rhett kisses her "with a savagery and a completeness that wiped out everything from her mind but the dark into which she was sinking and the lips on hers," and takes her into the bedroom. Next morning, Scarlett awakens to reflect. She feels "humbled" and "hurt" and "used brutally" — but "she had gloried in it."[10] Feminist literary critic Marilyn Friedman has complained that the scene romanticizes rape or something close to it. But Helen Taylor, author of *Scarlett's Women: "Gone with the Wind" and Its Female Fans*, conducted a survey of women who had seen the movie or read the book and found that "by far the majority" of them saw the staircase episode not as rape, but rather as "mutually pleasurable rough sex," a fantasy they felt was "erotically exciting, emotionally stirring and profoundly memorable."[11]

The football star is admired, even loved when he demonstrates his power and controlled violence on the field, but he is expected, somehow, to turn off those very qualities and abilities in every other situation. Should we be surprised when a champion hockey player turns out to be a rapist, when a great football player is shown to be a wife beater? Or should we instead be amazed at the human capacity for self-control, for channeling impulse and inclination that enables most other hockey and football stars to confine their terrific capacity for mayhem to the ice or astroturf? Many women would prefer it otherwise, but in the real world, the tough guy finds himself besieged with female admirers, while the self-effacing friend sadly clutches his glass of Chablis at the fern bar alone. The individual men and women who make up our species are extraordinarily ready to admire, to love, and to reward male demonism in many of its manifestations; and that admiration, love, and rewarding perpetuates the continuation, for generation after generation, of the demonic male within us.

Women don't ask for abuse. Women don't like many specific acts of demonic males. But paradoxically, many women do regularly find attractive the cluster of qualities and behaviors — successful aggression, dominance and displays of dominance — associated with male demonism. Both men and women are active participants in the very system that nurtures the continued success of demonic males; and the knot of human evolution, with the demonic male at center, requires an untying of both strands, male and female.

BREAKING FREE OF THE TRAP. Friedrich Engels regarded the historical institution of marriage as the beginning of the end for humankind: the start of the bourgeois family, patriarchy, and from there class and social struggle. For Engels, as for many traditional feminists, women are caught in a trap constructed by men — one marked with violence and perpetuated through a patriarchy of specific historic and social origins.

Evolutionary feminism provides a longer look. Human patriarchy has its beginning in the forest ape social world, a system based on males' social dominance and coercion of females. We can speculate that it was elaborated subsequently, perhaps in the woodland ape era, perhaps much later, by the development of sexual attachments with the same essential dynamic as gorilla bonds: women offering fidelity, men offering protection from harassment and violence by other men. From those poorly articulated forms of pairing, language would eventually generate both marriage and the patriarchal rules that favor married men. Men, following an evolutionary logic that benefits those who make the laws, would create legal systems that so often defined adultery as a crime for women, not for men — a social world that makes men freer than women.

Traditional feminism, even if it accepted the evolutionary timetable, would want to stop the analysis there and restrict the blame to men. But evolutionary feminism views women as active players in the development of the patriarchy, albeit often suffering as an ironic outcome of serving their own interests. Women's interests, their strategies and counterstrategies in response to men, have clearly played a major role in the development of human social forms.

Because we wanted to understand the source of the male aggressive strategies that have such a huge effect on ape and human societies, this book has been principally about males. As men, we have probably inadvertently neglected issues that women writers would have raised. But our focus on male aggression doesn't mean we imagine female strategies unimportant. It so happens that among chimpanzees, as among orangutans and gorillas, females have been unable to develop effective counterstrategies against those of their demonic male partners, though females with strong personalities will develop relatively satisfactory relationships with individual males. Among bonobos, though, females (newly freed from ecological constraints) responded effectively to the problem. The result, we have argued, was effectively a revolution in the nature of their society, the reduction of what once was an unpleasant form of patriarchy to a relatively tolerant and charming world in which the sexes are equal.

If there is no human group where women have achieved a comparable equality, women everywhere nevertheless have much of the same potential as bonobo females to change the system. Everywhere, women develop supportive social networks with each other. Everywhere, women wield some influence over their husbands and sons and other males — a power often much stronger than at first appears. The problem is that everywhere, women are caught in a trap. If they support each other too much, they become vulnerable to losing what they want, the investment and protection of the most desirable men. There is competition between women for the best men, and it can break the unwritten compact among women.

The interplay between women's interests in protecting themselves from abusive husbands on one hand, and finding or maintaining a long-term relationship on the other, is a classic tragedy of marriage. And all too often, despite the most appalling abuse, women don't leave a relationship. They stay because they are afraid, or because they forgive their mates, or because they hope to change them. Too often they stay because in some part of their mind, they want their mates, despite their aggression. We can take this as a metaphor for women's awkward place in the larger society. Individual women are caught in the trap of wanting a man to protect and provide for them; women as a group find their general interests ignored or stymied because some women side with the men.

Bonobos have shown us that the trap can be broken through female alliances. Among humans, a direct equivalent would be if women always remained together, day and night, in groups so large and well armed that they could always suppress the hostility of rowdy, aggressive men. The prospect seems too fantastic to consider further. Fortunately, humans can create other possibilities.

The problem in both human and ape history is that political power is built on physical power — and physical power is ultimately the power of violence or its threat. "Political power," in the words of Mao Tse-tung, "grows out of the barrel of a gun."[12] In other words, those who have political power can count on someone coming to their aid — the police or the military or the mob or the family, or the royal guard.

In traditional human societies political power is personalized. That is, it resides in the person (or families and alliances) of the most successful individuals and their descendants. Whenever political power is personalized, so is the physical power on which it ultimately depends; and whenever the physical power is personalized (not parsed and regulated through institutions, laws, and rules) the violence of demonic males from which it ultimately derives will be unrestrained. This is the broad current of human affairs, challenged in a thousand places by exceptional backwaters and eddies, but never reversed.

The alternative to personal is institutional. Italian politics, as Robert D. Putnam implies in *Making Democracy Work: Civic Traditions in Modern Italy* (1993), chose both routes after a collapse of medieval authority during the twelfth century, shifting toward institutional systems in the North while remaining traditionally personalized in the South. The result in this century? Southern politics is still profoundly corrupted by male power contests and stained by the secrets of Palermo. *

The great revolutions at the end of the eighteenth century in

---

* Our terms are necessarily abstract and theoretical. One might argue, for example, that the political system in the South of Italy was the development of an institutional monarchy, established in 1231 by Frederick II with a constitution formalizing the power of the king and his nobility. One might also note that the crucial system of the North, communal republicanism, established during the twelfth century by mutual aid societies, was marked by a personalized sort of civil engagement. Putnam indeed prefers to focus on the habits and traditions of civil engagement in the North and disengagement in the South; he sometimes describes the contrasting systems of power distribution as horizontal in the North and vertical in the South. Nevertheless, the North's communal republicanism fully relied upon institutional agreements to limit individual power, the creation of "elaborate legal codes . . . to confine the violence of the overmighty." The North's "institutions" ranged from solemn oaths guaranteeing mutual assistance to complex governmental bureaucracies to the invention of credit as a means of consolidating wealth. The power of paper — used to define

France and America marked a historic though incomplete shift from personal to institutional political power in those places. Once political power was taken away from individual men (who held it for life, by tradition and inheritance) and assigned by law and institution to temporarily appointed men, the physical power from which it derived also became, to a degree, regulated by institution. The underlying grip of male demonism, which attaches itself to human affairs through personalized physical power, was loosened slightly. The shift from personal to institutional actually describes not so much a single event as a process: a widening of the distribution of control away from individuals and cabals toward a more democratic spread. Democracy itself is a process; and its success requires long-standing habits and traditions of civic engagement, not merely the miraculous appearance of good constitutions or enlightened legal codes. Women in nineteenth-century America acquired the beginnings of political power through their own struggle, of course — but also because they lived within a system in which political power had been institutionalized to the degree that someone actually counted votes.

institutional relationships — was such that the city of Bologna alone, with a full population of about 50,000, required some 2,000 professional notaries. In the South, meanwhile, Frederick's constitution simply consolidated an already existing authority based on family and personal inheritance. Government was personalized, in our sense of the term, and over time, in the absence of a distributed, depersonalized system of institutional authority, family-based mini-governments such as the Mafia flourished. As one observer in Sicily noted in 1876: "Matters naturally reached a point where the instinct of self-preservation made everyone ensure the help of someone stronger; since no legitimate authority in fact existed, it fell to clientelism to provide the force which held society together. . . . A very unequal distribution of wealth; a total absence of the concept of equality before law; a predominance of individual power; the exclusively personal character of all social relations; all this [was] accompanied (as was inevitable) by the bitterest of hatreds, by a passion for revenge, by the idea that whoever did not provide justice for himself lacked honor."[13]

Institutions, even self-consciously democratic ones, are never static or perfect; and people's attachment to democratic institutions cannot be passive. The world's largest modern democracy, India, happens to rise over an ancient foundation that perpetuates, in the words of one informed commentator, "undoubtedly the most complex and rigid system of social hierarchy in the world."[14] And the Chipko environmental movement of the central Indian Himalayas, where men and women from a rural, traditional village culture joined across class and caste to challenge the destructive logging practices promoted by a formally democratic government, looks at first to be a reversal of the usual pattern, just as the relative social equality of Indian tribal society looks like a reversal of the subordination of women promoted by mainstream Indic and Islamic thought. But Chipko is a protest movement, not a governing one; and Indian democracy itself remains flexible enough to respond to Chipko and the numerous other grassroots movements that have suddenly appeared to resist the old ways.[15]

So personalized political power favors the appearance of the demonic male behaving in a male competitive style. The human political system most likely to favor a female competitive style is the one in which power has been depersonalized through the construction of stable institutions. Of the many styles of political institutions, the most depersonalized are also the most democratic. Among the nation-states, therefore, institutional democracies present the best actual situations where women can hope to acquire political codominance with men. Institutional democracies are deeply imperfect, in several instances weakened by civic and economic crises, often still highly personalized and therefore inevitably patriarchal. But they are also remarkably flexible and resilient, and a reduction in interpersonal violence seems most positively assured by faith in the evolution of institutional democracy.

In true institutional democracies, political power ultimately comes from the ballot box. And it is to the ballot box that women in the real world can most effectively mass themselves

— following the style of both Herlanders and bonobo females — and break through the trap defined by male interest. Feminist commentator Naomi Wolf has remarked on the peculiarity that women in democracies, who after all represent half the voters, have not learned to use their power more effectively. But the trend is there.

THE PROBLEM OF INTERGROUP AGGRESSION. On April 19, 1995, in downtown Oklahoma City, Oklahoma, two young men park a Ryder rental truck packed with a two-ton bomb in front of the Alfred P. Murrah Federal Building during working hours, a time when they can be certain the building will contain around five hundred ordinary adults of both sexes and, in a day care center on the second floor, about two dozen children and infants. The two men light a fuse and walk away from the truck. At 9:02 A.M. the bomb ignites and produces a blast powerful enough to disintegrate an entire side of the nine-story building, top to bottom, creating a giant orange fireball and a chaos of glass, concrete, steel, dust, office furniture, people, and body parts. The blast tears into the second-floor day care center and rips away toys and blocks and picture books along with children's faces, heads, arms, and fingers.[16]

Within days a prime suspect is captured, one Timothy McVeigh, an all-American male recently retired from the U.S. army and having some background connection with an informal group calling itself the Michigan Militia. McVeigh, standing soldier stiff and looking out through blue eyes and a lean, vacant face, appears to regard himself as a hero and prisoner of war, and refuses to say one word about himself, his ideas, his motives, or his possible association with the bombing. The Michigan Militia turns out to be the creation of a gun dealer named Norman Olson who believes the United Nations is poised to take over the United States and has therefore organized 12,000 gun-toting men across his home state to prepare resistance. Members of the group, who informally describe themselves as American patriots, describe seeing black helicopters flying overhead, obviously preparing the way for an international takeover. "Armed conflict may be necessary if the country doesn't turn around," Olson declares.[17] He later blames the Oklahoma explosion on the Japanese, while another spokesman for the group, Brigade Executive Director Stephen Bridges, righteously denies any association whatsoever with the bomb attack or its perpetrators.

The details are new: serious patriots and pot-bellied ideologues loyal to what they believe are original American ideals construct among themselves a vision of threatening forces as solid as the guns they carry and as fantastic, in its elaboration, as a visit from outer space. But the pattern is classic. As David Trochmann, cofounder of another American group of similar bent, the Militia of Montana, expresses the mindset: "Where we come from is very simple. There are good guys and bad guys out there. The bad guys have to be stopped."[18]

The system of thought, feeling, and behavior is no different in its dynamic and underlying psychology from that of a thousand other predominantly male groups, including urban gangs, motorcycle gangs, criminal organizations, pre-state warrior societies, and even the more formalized and state-sponsored armies (which after all still organize their fundamental fighting units at the platoon level). The psychology engaged may be hardly different from that expressed in predominantly male team sports — American football and hockey, for example. This behavior is familiar, not alien, and it reiterates a pattern as old and wide as the species. Demonic males gather in small, self-perpetuating, self-aggrandizing bands. They sight or invent an enemy "over there" — across the ridge, on the other side of the boundary, on the other side of a linguistic or social or political or ethnic or racial divide. The nature of the divide hardly seems to matter. What matters is the opportunity to engage in the vast and compelling drama of belonging to the gang, identifying the enemy, going on the patrol, participating in the attack.

An unblinking look in the mirror is a nerve-wracking experience: none of us likes to see our warts. But even if the sight is fearsome it at least brings the possibility of suggesting remedies.

Accepting that men have a vastly long history of violence implies that they have been temperamentally shaped to use violence effectively, and that they will therefore find it hard to stop. It is startling, perhaps, to recognize the absurdity of the system: one that works to benefit our genes rather than our conscious selves, and that inadvertently jeopardizes the fate of all our descendants. So does this study of our warts help at all? Does it help us take the step we would all like, to create a world where males are less violent than they are today?

It would be nice to answer yes, of course, but nothing suggests that a long view of the problem can seriously reduce the violence projected outward from a human society: the Us versus Them problem of intergroup aggression. Certainly at the international level one finds it hard to imagine how an evolutionary perspective could affect the calculations and aspirations of leaders under pressure to work on behalf of their own tribes or nations or empires. And history suggests that intellectual analysis has had vanishingly little impact on the course of intergroup aggression; as we survey societies from ancient Greece to modern-day nations we can detect no clear pattern in the overall rates of death from intergroup violence, which remain between 5 and 65 per 100,000 per year.[19] Every generation may hope that the big war is the last war, but so far there has been no sign that it is so. And while the human male temperament remains amazingly stable, human technology has suddenly, in the last historical instant, changed the stakes utterly. Our Pleistocene ancestors were beleaguered by their own demonic males, surely. But they didn't have automatic rifles, fertilizer bombs, dynamite, nerve gas, Stealth bombers, or nuclear weapons. We do, and therein lies the danger.

The trend in the scale of fighting offers a perspective perhaps more hopeful, though even this is fraught with alarming prospects. The groups that do the fighting have shifted in size from a handful of relatives in preagricultural days to the millions who fought in alliances in the present century. Though George Orwell imagined in *1984* that three world powers could coexist, each so powerful that no two would defeat the third, we find no real cases that suggest this sort of deadlock could happen in a stable fashion (which may not be such a bad thing, given the bleakness of Orwell's vision), so by simple extrapolation we expect alliance group sizes to continue rising. If they do, the ultimate effect might be to create a single world government in the near future, perhaps within a century or two or three. If we are lucky, it will be reached by agreement rather than by violence, and it will lead to a massive reduction in the frequency and scale of war.

The prospect of a single world power conjures up the excellent vision of a modern Pax Romana or the stability enforced by ancient Egypt or the thousand-year peace on Easter Island. But by the same token it alarms, not just because the central power might be morally malign, but also because it would doubtless play favorites. So the right-wing militias of the United States fear for their freedoms in a world led by the United Nations in much the way that Islamic militias of the Middle East do. Almost everyone thinks the United Nations is run by their enemies. In short, even if the world achieves political unity, fights can still be expected to continue, with demonic males strutting their stuff in the usual ways.

Traditionally, the most effective control of rogue behavior within societies has been moral sanction. In small-scale primitive societies, worship of an ancestor unifies descendants into common defense of the ancestor's honor. In larger groups, religions based on moral principle unite believers. Both systems work wonders within societies. But alas, they also almost invariably use their cooperative spirit against others, so that religious differences become fault lines for externally directed aggression. One of the merits of an evolutionary view is that it presents humans as a single group, worshiping, as it were, a single ancestor. It stresses our unity and trivializes our differences. The long view teases us with the thought that our foolish little status-seeking strategies, designed to increase our individual reproduction over a generation or two, could eliminate our species reproduction forever. If we could absorb the spiritual thought that humans — of all colors, creeds, sexes, and genders; residents and immigrants; conquerors and refugees — all of us, are descended from the same apes and that everyone's future depends on the abandonment of imperialism, we might come to think that a rise in status is less important than the protection of peace, and to be more generous to our rivals. But these are distant and impractical thoughts when it comes to solving immediate, practical problems, where the most powerful players' hearts too often burn with their inner desire to be alpha males. The growth of world moral systems may eventually affect politics as strongly as local religions have, but not for a long time.

In the shorter term, the remedies for male violence occupy the domain of politics, not biological philosophy. But as evolutionary biologists, we can at least stake out the intellectual battleground. We should accept the likelihood that male violence and male dominance over women have long been a part of our history. But with an evolutionary perspective we can firmly reject the pessimists who say it has to stay that way. Male demonism is not inevitable. Its expression has evolved in other animals, it varies across human societies, and it has changed in history. Natural selection makes it inevitable for each individual to pursue his or her own interests, and for conflicts to arise and be resolved. In human affairs conflicts have traditionally been resolved in favor of high-status men because they were able to control power so effectively. But the nature of power, its distribution and effects and ease of monopolization, all depend on circumstance. Add to the equation some of the more obvious unknowns, such as the democratization of the world, drastic changes in weaponry, and explosive revolutions in communication, and the possibilities quickly expand in all directions. We can have no idea how far the wave of history may sweep us from our rougher past.

13

$'$ $'$ $'$

KAKAMA'S DOLL

ON A SHELF IN MY OFFICE rests an undistinguished piece of wood the size of an airline pillow. Half-rotted before it dried, it bears the typical scars of forest life: holes bored by beetle larvae, a tear from a long-gone branch, cuts and scratches from the jostlings of passing animals. It's an ordinary piece of an ordinary tree, just like hundreds of other such fragments that could be found during a half-hour walk in Uganda's Kibale Forest.

Kakama was two years old when I arrived at Kibale in 1987. His mother had been given the name Kabarole, and Kakama was her firstborn. The mother, having lost her right hand to a poacher's snare, was only too easy to recognize, so when we began our study we were always able to identify her and, consequently, we observed Kakama's childhood more closely than any other young chimpanzee in the forest.

After Kakama was born, it was six years before his mother resumed mating. She didn't become pregnant for another two years. Then, during the first week of February 1993, in a flurry of male mate-guarding and attacks and falls and chases and screams, Kabarole mated a hundred and twenty times with a dozen males, and Kakama's first sibling was conceived. I was there, but who the father was I couldn't say.

One day in May of 1993 I spent a quiet morning alone with Kakama and his pregnant mother as they moved between fruit trees. Kakama was full of his normal bounce, out of step with Kabarole's slow mood. Twice, while the mother paused for a few minutes during her walk to the next tree, the son tried to get a rise out of me. He stamped the ground and displayed toward me for a few steps before slapping a tree and retreating. As usual, I continued to sit and groom the hairs on my arm. I wanted to be boring.

The second time he did this, he somersaulted away and came to rest straddling a small log. For a moment he lay face down with his tummy on the log, then he continued his roll with the log held to his belly. Two more turns and he stood up, holding a stub on one end of the log with his right hand. His mother was up and walking down the slope. Kakama walked after her, dragging the piece of wood behind him like Christopher Robin with Winnie-the-Pooh. *Bump bump bump.*

The vegetation grew thicker, and I lost the chimpanzees for a few minutes. *A pity,* I thought. *I'd have liked to see what Kakama did with that log.* But I was lucky. I heard climbing, and soon I found myself below a towering drypetes tree with fresh ripe fruits fallen on the ground. It took me ten minutes to find a hole in the canopy and see that Kabarole and Kakama were up there, feeding alone. Little Kakama still had his log with him. While he fed, the log lay next to him. When he moved, he picked the log up. He took it with him wherever he went, perhaps five or ten moves. Then it was rest time.

I watched Kabarole make her nest first, but by the time I found Kakama he had already made his and had been sacked out in it for several minutes. All I could make out were his hands and feet. He was lying on his back with his limbs in the air, and through my binoculars I could see that he was holding the log over him. Just as some mothers do with their babies, he slowly juggled and balanced the piece of wood with his hands and feet.

Then his limbs and the wood disappeared into the nest, and I saw no more movement for a while.

Half an hour later Kabarole left her nest, and then Kakama left his, too. Like me, he must have expected Kabarole to lead onward to the next tree, but Kabarole just sat in the high *drypetes* and looked out over the forest.

So Kakama sat for some time, too, with his log next to him. After a while he picked it up and climbed a few meters. It was awkward, climbing while carrying an object half his length. Then he made a new nest rather quicker than usual, in about half a minute. Completing it, he put the log into the nest, and sat next to the nest. Two minutes more, and he climbed in, too, and disappeared from view.

I was with Kakama and his mother for another two hours. They visited two more trees, and Kakama kept the thing with him at all times but once. The exception occurred when the log fell while Kakama was eating leaves thirty feet up in a *chaetacme* tree. He watched it land and resumed feeding. *So that's the end of it*, I thought. But when he followed Kabarole out of the tree he immediately went down the slope to collect his log before turning back to follow his mother.

During the course of the morning, Kakama carried that piece of wood in every manner imaginable. He walked upright with it on his back, cupped in two hands or held in one. He walked tripedally with it pressed against a thigh or dragged on the ground. He carried it on the back of his neck. To do that wasn't easy, but he was persistent.

After noon, and after they had visited the fourth food tree of the day, Kabarole led Kakama and his log into a thick swamp where their trail became elusive. I hoped for helpful sounds, but I ran into the snuffling and threatening snorts of a bush pig. By the time I had skirted those signs of danger, the chimps and that piece of wood had vanished.

My intuition suggested a possibility that I was reluctant, as a professionally skeptical scientist, to accept on the basis of a single observation: that I had just watched a young male chimpanzee invent and then play with a doll in possible anticipation of his mother giving birth. A doll! The concept was novel enough that I simply filed away my notes without saying much about it to anyone else, and left Uganda the following week.

Four months later, two field assistants at Kibale, Elisha Karwani and Peter Tuhairwe, happened to be following Kabarole and Kakama. Neither Karwani nor Tuhairwe knew of my observation. Yet for three hours they watched Kakama carry a log — not the same one as before, surely — taking it with him wherever he fed. This time they saw him leave it. Once they were certain Kakama had disappeared, they collected it, brought it to camp, and stapled to it a label that described their own straightforward interpretation of the object's meaning: "Kakama's Toy Baby." Five weeks later, Kabarole gave birth to a healthy daughter, Omugu.

Both sets of observers thought of Kakama's logs as dolls merely from the way he carried them, but the clearest sign that Kakama treated the logs as imaginary babies was the time he made a nest and put the log into the nest on its own. Kakama was exactly the sort of youngster who might be expected to want a play partner most. He was an only child with a playful personality, his mother was relatively antisocial, and she was now pregnant. Could he have known that his mother was pregnant and therefore looked forward to a sibling so intensely that he created one in his mind?

Those of us who know the great apes well often get into trouble with our scientific colleagues because we see evidence of mental events beyond what's easily provable. Francine Patterson used sign language to ask the language-trained gorilla Koko to paint pictures representing two emotions, love and hate. Patterson asked me to guess which painting was which. One presented a twirl of soft reds. The other was full of sharp angles painted in black. It was obvious. But it's hard to *prove* that Koko even understood the instructions, let alone that her paintings represented emotional states like those felt by humans. Many years earlier, Cathy Hayes watched her adopted chimpanzee, Viki, drag an imaginary pull-toy behind her. The imaginary string got imaginarily stuck when it was wound too tightly round a potty, so Viki had to have Hayes's help to unwind it.[1] Try selling that to your neighborhood skeptic and you'll see not only how hard it is to convince a doubter but also how difficult it is to explain certain things about the lives of apes without resorting to the concept of imagination.

Apes are caught between two worlds, of human and nonhuman consciousness. Ape observers are caught between two parallel worlds, between being convinced of apes' mental complexities and finding them hard to prove. The rules of proof have yet to be written for unreplicable phenomena that seem to require a ridiculous number of coincidences or rare events to be explained in any way other than as a result of complex cognition.[2] Even if we can't prove claims about what apes know and think and feel, though, we will make more mistakes by ignoring such signs of mental power than by taking them seriously. With apes, too many intriguing stories suggest that there are minds in the forest.

Animal complexity and intelligence have increased steadily in the 4 billion years of life's history. One billion years ago the cleverest species was some unknown microscopic bit of goo. At 100 million, a fish perhaps, or an early mammal. At 10 million, an ape or a dolphin. At a million, early humans, maybe already on the verge of using some simple form of language.

In another world, where the Pliocene climatic extremes were a little gentler, where an asteroid didn't suffocate most of life just when it did, where gravity was stronger or continents more mobile or water more abundant, the particular species with the biggest brains at any time would have been different. But in most lines brains would have gotten bigger with time, just as they have in our own little life system, not just in the human line, but also in fish and lizards and birds and carnivores, in elephants as well as in dolphins.[3] Eventually, they would have enlarged enough to allow a rudimentary form of language. Chimpanzees and bonobos are almost there — to judge from individuals like Kanzi, the famous bonobo living in an American laboratory who can understand the spoken word so well.[4] Dolphins may be almost there. One hundred fifty thousand years ago our ancestors were there or almost there. And then they crossed the line; humans entered a new conceptual world where ideas could be shared and the future discussed and other species scorned. So what would happen in our slightly different universe, where the asteroids were smaller or the earth's core colder or the year shorter? Brain size would still increase, pushed by the arms race of predator and prey or by the need for outsmarting others of your own species. And just as it happened in our own planet's history, several different species would approach the language barrier. Who would get to cross that barrier first this time around? It could be something like a human ape again. But it might equally well be a monkey-like ape or a hyena or a manta ray or a parrot or a velociraptor. Whichever species crossed the language barrier first might have had demonic males or demonic females or pacifists in either sex or, for that matter, monogamous pairs locked in isolation within distant, birdlike territories.

The linkage of high brainpower to male demonism looks like a tragic coincidence of independent causal chains; but there's more to the connection than that. Intelligent minds are responsible for new forms of aggression irrelevant to animals without good memories and long-term social relationships. And relationship violence is not the only outcome of a large-brained demonic mind. Political complexity creates coalitions, which can lead to the massive imbalances of power that foster violence. Ingenuity now serves the demon with new weapons, new tactics, new kinds of deception in the ever-escalating game of conflict.

For us, the biggest danger is not that demonic males are the rule in our species. After all, other demonic male species are not endangered at their own hands. The real danger is that our spe-

cies combines demonic males with a burning intelligence —
and therefore a capacity for creation and destruction without
precedent. That great human brain is nature's most frightening
product.

But it is simultaneously nature's best, most hopeful gift. If
we are cursed with a demonic male temperament and a Machiavellian capacity to express it, we are also blessed with an intelligence that can, through the acquisition of wisdom, draw us
away from the 5-million-year stain of our ape past. Intelligence
is something we are familiar with, an old book, an old friend. But
what is wisdom? If intelligence is the ability to speak, wisdom is
the capacity to listen. If intelligence is the ability to see, wisdom
is the capacity to see far. If intelligence is an eye, wisdom is a
telescope. Wisdom represents the capacity to leave the island of
our own selves and to move out across the sea. To see ourselves
perhaps as others do, and to see others within and beyond the
first dimension or context: of time and space and being. Wisdom, in other words, is perspective.

Temperament tells us what we care about. Intelligence helps
generate options. And wisdom can bring us to consider out-
comes distantly, for ourselves and our children and our children's children . . . and perhaps even for the minds in the forest.

FAMILY TREES

, , ,

MAPS

, , ,

NOTES

, , ,

BIBLIOGRAPHY

, , ,

ACKNOWLEDGMENTS

, , ,

INDEX

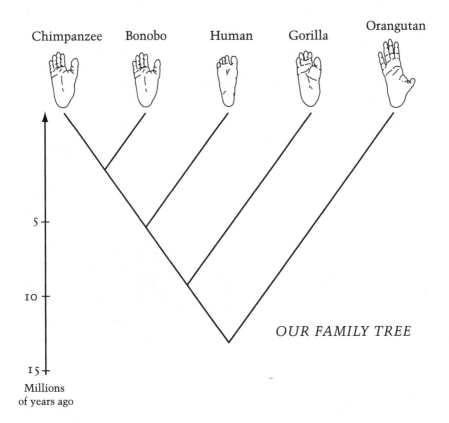

OUR FAMILY TREE

5

10

15

Millions
of years ago

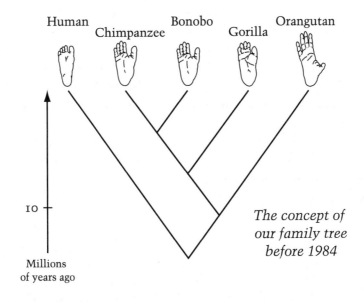

10

Millions
of years ago

*The concept of
our family tree
before 1984*

AFRICAN APE DISTRIBUTION MAP

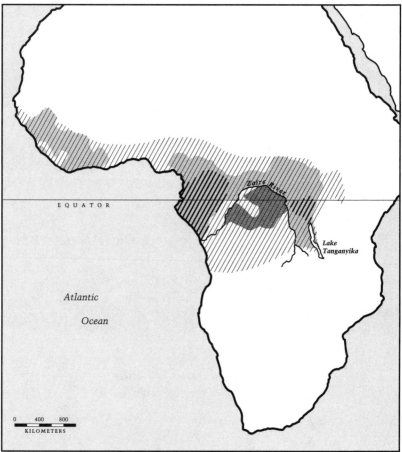

EQUATOR

Zaire River

Lake
Tanganyika

Atlantic

Ocean

0   400   800
KILOMETERS

Rainforests (and transitional edge)

Chimpanzees

Gorillas and Chimpanzees

Bonobos

SOME PRINCIPAL APE RESEARCH SITES
AND OTHER POINTS OF REFERENCE

SENEGAL

Niokolo-Koba

Bossou

Taï Forest   IVORY COAST

UGANDA

Zaïre River

Djolu

Kibale Forest
Kampala

EQUATOR

Wamba

Lake Victoria

Bukavu

ZAÏRE

Mahale Mountains

Bujumbura

Gombe Stream

Lake Tanganyika

TANZANIA

Atlantic

Ocean

0   400   800
KILOMETERS

● Town or village

▲ National park or protected forest

⊙ Capital city

○ Research site

NOTES

I. PARADISE LOST

1. The quotation is from Watson (1994b): 27. See also "Burundi" (1993); Shoumatoff (1994); and Watson (1994a).
2. Atrocities in Rwanda, see Fritz (1994); Lang (1994).
3. Quotation by a "reporter for *Newsweek*" from Hammer (1994): 34. Quotation from Ngoga Murumba: Lorch (1994): 1.
4. Bauman (1926).
5. See McGrew (1992) and Wrangham and others (1994). Use of leaf-cushions was recently described by Rosalind Alp, from observations in Outamba-Kilimi National Park, Sierra Leone.
6. Ardrey (1966): 222.
7. This and the subsequent discussion of the discovery of intercommunity violence at Gombe is largely based on Goodall (1986): 503 ff; Goodall and others (1979); and Wrangham (1975).
8. Power (1991).
9. Nishida, Haraiwa-Hasegawa, and Takahata (1985). In 1985 it was thought that all the males had died. Later one male was found still alive, living alone in the area occupied by the former K-group.
10. Brewer (1978).
11. Boesch and Boesch (1989): 567.

12. Morell (1995).
13. Reynolds (1967).
14. Cartmill (1994); Lorenz (1966); Eibl-Eibesfeldt (1989): 406.
15. Bigelow (1969); Tiger (1969); Tiger (1987); Alexander (1987).
16. Matsuzawa (1985).
17. Rodseth and others (1991); Manson and Wrangham (1991).
18. This generalization is debated within anthropology. See Knauft (1991) vs. Rodseth and others (1991).

2. TIME MACHINE

1. See for example the creationist perspective presented by Gish (1978) and Parker (1980). They have been well evaluated by Eldredge (1982).
2. Hunt (1994) provides an analysis of structure and function of the hominid shoulder.
3. Based on RWW's experience of talking to Zaïrois. The first live chimpanzee reached Europe in 1640, though rumors of the species' existence preceded its arrival. See Peterson and Goodall (1993): 15–17. The first European scientist to claim that humans are primates was the great Swedish classifier, Carolus Linnaeus, who during the eighteenth century established the modern system for classifying plants and animals, had apparently seen a chimpanzee, and placed humans into his Order of Primates (monkeys, apes, and prosimians).
4. Darwin (1871): 190.
5. This and subsequent Darwin quotations from Darwin (1871): 191, 198, 194–199.
6. This and subsequent Huxley quotation from Huxley (1863; 1894): 97, 92. Discovery of gorillas: See Reynolds (1967). Also Short (1980). Gould (1995) places Huxley's interest in gorillas in the context of "the great hippocampus debate" with Richard Owen.
7. Pilbeam (1996).
8. See Gould (1995).
9. Keirans (1984).
10. Nuttall (1904).
11. See Dennett (1995): 68–73 for an elaboration of this argument.
12. By 1933 enough striking similarities had been identified to justify British zoologist Solly Zuckerman placing, in his highly influential

book *Functional Affinities of Man, Monkeys, and Apes,* humans right next to the great apes in the tree of scientific classification.

13. Some possibilities put forward: Schwarz (1984); Zihlman and others (1978).
14. Zuckerman (1933): 178.
15. Goodman (1963).
16. Sibley and Ahlquist (1983).
17. Sibley and Ahlquist (1984).
18. Sibley, Comstock, and Ahlquist (1990).
19. Caccone and Powell (1989). The critic was Jonathan Marks, e.g., Marks, Schmid, and Sarich (1988).
20. Analyses of Nuclear DNA: (1) β-globin gene: Williams and Goodman (1989); Bailey and others (1992); (2) Immunoglobulin pseudogene: Ueda and others (1989); (3) Ribosomal gene: Gonzales and others (1990); (4) alpha-1,3-galactosyltransferase gene: Galili and Swanson (1991); (5) Mitochondrial DNA, cytochrome oxidase II gene: Ruvolo and others (1991); (6) Mitochondrial DNA for 6 proteins, 11 transfer RNAs: Horai and others (1992). As the data on different segments of genetic material accumulated, only one exception to the rule of chimpanzees being closer to humans than to gorillas appeared. This was the involucrin gene, which evolves too fast, and therefore accumulates too many changes, to be used as a marker of ancestral relationships. See Djian and Green (1989). For problems with using the involucrin gene, see Bailey and others (1992). Marks (1993) is a defender of the old view.
21. Eldredge (1982): 92–94.
22. Information from nucleotides in the mitochodrial genome: Horai and others (1995). Nuclear DNA sequences: Takahata (1995). Several billion nucleotides of the entire nuclear genome: David Pilbeam, personal communication (1995).
23. Lewin (1993).
24. The story of the kooloo-kamba is based on Shea (1984); also Short (1980). Quotations of du Chaillu are taken from Shea (1984): 2; and Short (1980): 5. The other quotation is from Short (1980): 6.
25. Shea (1983). See page 55: "Compared to the chimpanzee, the gorilla's low level of encephalization, small neonate, craniofacial morphology, strong sexual dimorphism, and overall growth patterns all support the hypothesis that differential concentration of similar growth patterns during postnatal development primarily distinguish the gorilla from its presumably Pan-like ancestors."

26. Some anatomists have detected sufficient similarities between humans and chimpanzees to support their close genetic relationship on morphological grounds. See Groves (1988).

27. White, Suwa, and Asfaw (1994); WoldeGabriel and others (1994). Body weights for *Australopithecus ramidus* have not yet been carefully worked out, but this species was close in size to *A. afarensis*, which is thought to have weighed about the same as chimpanzees, i.e., 29 kilograms for females, 45 kilograms for males (McHenry [1994]). Chimpanzee weights in Gombe average 29.8 kilograms for females and 39.5 kilograms for males (Wrangham and Smuts [1980]). In Kibale and some other parts of Africa male chimpanzees can average 48 kilograms or more, with females also larger than in Gombe.

28. White, Suwa, and Asfaw (1995).

29. Wood (1994).

30. Darwin (1871): 199.

31. Evidence for the apes being morphologically conservative comes from estimates of how long their subspecies have been separate. For example, the divergence of western and central subspecies of chimpanzees (which are very similar morphologically) is calculated from genetic data at 1.58 million years. This suggests that very little morphological change has happened during the last 1.58 million years. See Morin and others (1994).

32. For a summary of the evidence linking an asteroid impact to mass extinction, see Kerr (1992).

3. ROOTS

1. For an overview of the ecological background to hominid evolution, see Foley (1987) or the *Cambridge Encyclopedia of Human Evolution* (Jones, Martin, and Pilbeam [1992]). The specific problem of hominid origins was discussed by Coppens (1994). Like ourselves and most researchers, Coppens assumes that hominids evolved as a result of a drying event. But we disagree with Coppens about what caused the drought. Like Kortlandt (1972), Coppens suggests that the event that created newly dry environments was the emergence of the great East African Rift Valley. However, this happened at 8 million years ago, somewhat too early; it suggests that austra-

lopithecines would be confined to areas east of the Rift Valley, whereas they are now known from farther west, in Chad (Brunet and others, 1995); and it implies that the Rift Valley would be a major barrier, which is speculative at best. Climatic changes such as the one portrayed in the present chapter happened much more frequently than major geological events and are all that is needed to explain how forest apes would have been isolated in drying woodlands.

2. Hunt (1994).

3. The inference that bipedal locomotion led to longer daily travel distances comes from comparing chimpanzees and foraging people. For instance, among Gombe chimpanzees, average distances are about 3.0 kilometers daily for mothers, versus 4.9 kilometers for males (Goodall, 1986). Travel distances for Kanyawara chimpanzees appear to be very similar. Efé Pygmy men walked 9.4 kilometers per day (Bailey, 1991). Bipedal locomotion appears to allow faster walking than knuckle-walking, e.g., 3.4 kilometers per hour (Efé men, Bailey 1991) versus 2.1–2.8 kilometers per hour (chimpanzees, Hunt 1994).

4. *Agriotherium* was a bear reported in Ethiopia and well-described from South Africa. It appears to have been a tropical equivalent of a polar bear — fast, tall, and almost wholly carnivorous. See Wolde-Gabriel and others (1994).

5. Moore (1992).

6. Recent evidence on the distribution and seasonal availability of seeds, nuts, and other foods was reviewed by Charles Peters and Eileen O'Brien (1994).

7. Meat eating may have started as scavenging. See Blumenschine and Cavallo (1992).

8. In general, edible roots occur at low density in rainforests. The highest densities recorded were those found by Marcel and Annette Hladik (1990) in the Central African Republic, where they found tuber densities low inside the rainforest but reaching as high as 100 kilograms per square kilometer toward the forest edge. However, even at this density, tubers were in short supply for the local people according to Serge Bahuchet (1990), and even the highest forest figure pales beside the estimates of 40,000 kilograms per square kilometer found by Anne Vincent (1985) in a Tanzanian savanna woodland.

9. See Brain (1988). Susman (1988) discusses tool use by the robust australopithecines, and describes evidence that their hands were specifically adapted to tool use.

10. Hatley and Kappelman (1980) proposed the similarity between pig and hominid teeth.

11. The distribution of the major family of African mole rats (Bathyergidae) is shown by Jarvis and others (1994). The other sub-Saharan family is the Rhizomyidae, confined to high-altitude grasslands. The relationship between rainforest apes, hominid apes, and mole rats has not yet been formally studied, but is the subject of current work by Greg Laden and Richard Wrangham.

12. RWW visited in 1991 with Peter Howard and saw moss sponging but not root digging.

13. Wrangham (1981) describes the effects of water shortage for vervet monkeys.

14. Terèse Hart, personal communication.

15. Jones, Martin, and Pilbeam (1992).

16. We must stress that this is still a hypothesis. There is no consensus about the hominid diets. Promising methods of testing hypotheses involve direct study of fossils by analyzing mineral content or by looking at the microwear on the tooth surfaces. To date, however, the results are rather confusing, though they tend to favor a mixed diet including some meat. See, for example, Sillen (1992); and Walker (1981).

17. One line developed huge teeth, suggestive of massive commitment to low-quality foods such as roots. See Grine (1988).

18. Jones, Martin, and Pilbeam (1992).

19. The most convincing proposal for why brains expanded invokes a new feeding adaptation. Aiello and Wheeler (1995) note that brain tissue uses energy at particularly high rates, that overall metabolic rate is no higher in large-brained than in small-brained species, and that therefore, species can only afford large brains if some other organ is small. They show that the only other organ system sufficiently variable to allow some species to have larger brains is the gut; and that species eating easily digested foods are the only ones to have small guts. Therefore, by enabling guts to be small, an adaptation to high-quality, easily digested foods allows brains to be large. Among primates, species with relatively small guts indeed have relatively larger brains than those with big guts. For humans, a

massive increase in meat eating could therefore help account for brain expansion.

4. RAIDING

1. See Manson and Wrangham (1991): 370; also Turney-High (1949; 1991); and Otterbein (1970). A wonderful example of the cultural dimension of war comes from the Salu Mambi headhunters of highland Sulawesi (formerly the Celebes). Annual headhunts continue to the present day, with small bands making their traditional expeditions downstream in search of severed heads which, when brought back to the home village, will be welcomed for their effects in renewing the fertility of their fields and the prosperity of their households. Traditionally, the heads are apparently vital to the negotiations that maintain power balances among rival groups, so war has complex dimensions. That this does not depend on killing per se is shown by the modern form of headhunting: in deference to modernity, the raiders set out without weapons and return from their expeditions not with a real head, but with a coconut bought from a nearby town. See George (1991). For war as population control, see Harris (1979):90–92. For anthropological definitions of warfare compared to chimpanzee intercommunity aggression, see Boehm (1992). For a superb overview of some general issues on the prehistory of war and an excellent, up-to-date assemblage of the anthropological and archaeological evidence, see Keeley (1996).

2. Howard (1983): 7.

3. Material on the Yanomamö here and following based on Chagnon (1988; 1992).

4. See Albert (1989; 1990); Chagnon (1990).

5. Chagnon (1992): 5.

6. See Kuper (1994): 144.

7. Chagnon (1988): 989.

8. This section and following based on Manson and Wrangham (1991).

9. Calculated from data presented by Goodall (1986): 110. The cause of death was known for only fourteen males. Five were the Kahama deaths (Sniff, Godi, Dé, Charlie, Goliath); nine died from disease (David, Leakey, William, JB, McGregor), old age (Mike, Hugo), or accidental injury (Rix, Huxley). Of the remaining seven, it is prob-

able that several died from intercommunity violence. This includes both Kahama males (Willy-Wally and Hugh, who disappeared during the period when other Kahama males were attacked) and Kasekela males (Sherry, Faben, Figan, and Humphrey, who disappeared during the period 1975–1982, when Kalande males were raiding into the Kasekela range). The estimate of mortality from aggression therefore ranges from 5 out of 21 (23.8 percent) to 11 out of 21 (52.4 percent).

10. Morgan (1852; 1979).
11. Turney-High (1949; 1991): 112.
12. Turney-High (1949; 1991): 23.
13. Turney-High (1949; 1991): 112.
14. See Manson and Wrangham (1991), note 3; also Ember (1978); and Otterbein (1970).
15. Meggitt (1977):1.
16. Knauft (1991).
17. Knauft (1991): 391.
18. Ember (1978).
19. Nance (1975).
20. Headland (1992).
21. Eibl-Eibesfeldt (1989): 409.
22. Daly and Wilson (1988).
23. Chagnon (1988): 986. Also Eibl-Eibesfeldt (1989): 417.
24. Chagnon (1988): 986.
25. Quote from Robarchek and Robarchek (1992): 197. Our entire account of the Waorani is based on this article.
26. Robarchek and Robarchek (1992): 205.
27. Otterbein (1970): 20, 21.
28. Robarchek and Robarchek (1992): 192.

5. PARADISE IMAGINED

1. See Marx (1964), especially Chapter 2 (34–72).
2. Gauguin's letter to his agent will be found in Guérin (1974; 1978): 159, 160. The painter had just asked his three questions in the context of a written attack on Catholicism, finally published as "The Catholic Church and Modern Times." See Guérin (1974; 1978): 161–173.

3. Further information on Gauguin and the painting comes from Cachin (1989; 1992).

4. The quoted comments from *Noa Noa* are in Guérin (1974; 1978): 83, 84, and 80.

5. Gauguin's reference to life in the Marquesas is from "Scattered Notes" in Guérin (1974, 1978): 274.

6. For background on the success of *Typee* and problems with publishers and potential publishers, see the Introduction by George Woodcock to Melville (1846; 1972).

7. Walter T. Herbert, Jr. (1980) has provided the best and most complete analysis of the the Marquesans and their place in Melville's imagination. We have relied on Herbert's scholarship throughout this section, starting with the cultural and historical background briefly surveyed here and moving on to the theory that Porter, Stewart, and Melville were examining the Marquesans through the cultural filters of Enlightenment rationalism, Calvinism, and Romanticism.

8. Reverend Stewart as quoted by Herbert (1980): 63.

9. Records show Melville deserted ship on July 9 and took passage on another ship on August 9. Allowing for travel time into Typee valley from Nukuheva harbor would take a few days, up to a week.

10. The quotation from Melville's Preface: Melville (1846; 1972): 34.

11. Who are the real savages? Tommo's diatribe on this matter appears in Melville (1846; 1972): 180. Biblical paradise: 265. Happy innocence, artless simplicity, and good-natured laziness are suggested on pages 211, 253, 236. Lack of predatory beasts or mosquitoes mentioned on pages 286, 285. The quoted remark about battle wounds is from page 247. Material on the "nymphs" and Fayaway will be found on pages 188–192, 133–136, and elsewhere.

12. Tommo's defense of cannibalism is from Melville (1846; 1972): 278. Tommo's desperate assault on the Typee warrior will be found on page 332.

13. As quoted in Bowlby (1990): 170.

14. We rely on Derek Freeman's *Margaret Mead and Samoa* (1983) as well as Adam Kuper's *The Chosen Primate* (1994) for some of the background discussion of the nature-nurture debate; the subsequent critique of Mead's work in Samoa is based on Freeman's research and tempered by reference to Kuper's discussion. The three Galton quotations are from Freeman (1983): 7, 10, 15.

15. Boas's statement of a "fundamental need" in Boas (1924): 164. Boas may have already imagined that adolescence would provide an ideal focus for such an investigation. Before he came to Columbia, the anthropologist had taught for a time at Clark University in Massachusetts until he resigned after a bitter argument with its founder G. Stanley Hall. In an ambitious 1907 book, *Adolescence: Its Psychology and Its Relations to Physiology, Anthropology, Sociology, Sex, Crime, Religion and Education*, Hall had promoted the theory that the stages of any person's life recapitulate the stages of human culture: from infantile savagery to civilized maturity. The universal agony and ecstasy of adolescence, then, was equivalent to a predictable moment of transition in the march of cultural progress, "suggestive of some ancient period of storm and stress when old moorings were broken and a higher level attained." Hall's claim offered a way to probe the nature-nurture problem. The presumption of a universal human adolescence, a time of inevitable "storm and stress" produced by glandular secretions or some comparable biological machinery, had to be challenged. See Kuper (1994): 180–182.
16. Mead's rhetorical question is from Mead (1928a): 11.
17. The details of her entry into Samoa and the cultural context there are based on Freeman (1982): 65 ff.
18. Mead's quoted insistence on close contact with her subjects: Mead (1928a): 10.
19. Her difficulties with learning the language are summarized in Freeman (1982): 65. In *Letters from the Field, 1925–1975* (1977): 29, Mead describes her fears about living with Samoans.
20. Mead mentions the ease of mastering "the fundamental structure of a primitive society" in Mead (1928a): 8.
21. Summary comparisons of Western and Samoan adolescence are from Mead (1928a): 234 and 38.
22. Comments about the ease of Samoan life and its sexual freedom: Mead (1928a): 198, 201. "No room for guilt": Mead (1940): 96.
23. Remarkable elimination of many psychological problems, mentioned in Mead (1928a): 243, 206, 106, 213, 206, 207, 215, 223 — and "practically no suicide": in Mead (1928b): 487.
24. Her comments on rape are from Mead (1928a): 93; and Mead (1928b): 487. Her remarks on sleep crawling: Mead (1928a): 93 ff.
25. Lack of violence in general: Mead (1928a): 198, 199. And as quoted in Freeman (1983): 90.
26. See Stocking (1989): 246.

27. Boas's foreword: Mead (1928a): xv.
28. Her 1961 remarks are quoted by Freeman (1983): 106.
29. See, for example, Levy (1983): 829. See Stocking (1989): 253, 254.
30. Stocking (1989): 257. See Stocking for a balanced view of the controversy generated by Freeman's attack.
31. Kuper (1994): 193.
32. Levy (1983). Another complaint against Freeman: having come to Samoa fourteen years after Mead, he failed to account for important historical and cultural changes that could have taken place during that time and so "seems unaware" of the significance of "historicity" (Leacock, 1993: 351). This criticism is not so convincing. The fact is that Freeman remains an authoritative source whose own portrait of the islands reasonably considers and tries to accommodate the vexing issue of historical change.
33. Mead's comments on her "deviants" and "delinquents" are from Mead (1928a): 169, 172. Freeman's assessments are from Freeman (1983): 93 and 258.
34. Data on ratio of male to female first offenders: Freeman (1983): 258, 259.
35. Virgins: Mead (1928a): 151. Of the twenty-five, eleven had had at least one heterosexual experience. Although ceremonial virgins were chosen from aristocratic families and given special status, virginity at marriage was through the *taupou* culturally idealized for Samoans of every rank, according to Freeman (1983): 227.
36. Information on Samoan adultery, suicides, rape, and surreptitious rape are provided by Freeman (1983): 104, 220–222, 243–249.
37. On war and its traditions in Samoa, see Freeman (1983): 157–173. Assault and murder rates are mentioned by Freeman on page 164.
38. Mead's apotheosis is described in much greater detail by Freeman (1983): 106, 107.
39. "Gingrich" (1995).

6. A QUESTION OF TEMPERAMENT

1. Dahomey women warriors are mentioned in Harris (1989): 285. For the best consideration of the Dahomey Amazons, see Law (1993).
2. As quoted in Law (1993): 252.
3. Law (1993): 258.
4. Discussion of the formalized male monopoly on warfare is largely

based on Adams (1983); quotations from 201, 202. According to Adams, aside from the facts that pregnant and nursing women might abstain from the demands of war or hunting, there is little reason why women are less "biologically" inclined to warfare. He cites his discovery of a strong correlation between institutionalized marriage and warfare patterns (all nine of the cultures that have women warriors also have exclusively "external warfare" and/or endogamy) as evidence for the "cultural construction" of war. While it is convincing that wives might be excluded absolutely from making war in societies where war may be fought against their fathers and brothers, Adams ignores the much larger patterns: no society excludes men; no society includes women except rarely or peripherally. He does hypothesize that exogamy may have begun to avoid primitive feuds, since "most primitive feuds stem from fights over women." But why, if these behaviors are part of the culturally constructed house of cards, can't we assume an equal number of primitive feuds over men?

5. Mentioned in Harris (1989): 278. See also Gray and Wolfe (1980); and Percival and Quinkert (1987).

6. Tuten (1982).

7. Griesse and Stites (1982): 74.

8. Amrane (1982).

9. Bloom (1982).

10. Statistics from Federal Bureau of Investigation (1991).

11. Quotation from Weil (1994); see Wolf (1993): 228–232, for a more serious discussion of the issue.

12. For example, Pagnozzi (1994); Pierre-Pierre (1995); and MacDonald (1991).

13. Adler (1975) argues for an increase in women's rates of violent crime six or seven times more rapid than the increase in men's; in fact, the male-female proportions for violent crime remain steady. Simon (1975) focuses on property crime, where there has been a proportionate increase, particularly in areas largely traditional for female criminals to begin with, such as petty theft, fraud, and vandalism. Simon, in fact, noted "large increases" in property crimes and "the absence of increases" in murder, assault, and "other violent crimes" (47). For a summary of these issues, see Flowers (1989): 85–87. See also Leonard (1982): 27 for a reasonable assessment: The apparent increase in percentages of girls and women arrested for serious

crimes between 1955 and 1970 results from the combination of some increase in serious property crimes with no proportionate increase in crimes of violence. The issue of "girl gangs" deserves separate consideration. The Lady Eights of San Antonio has been described as specifically associated with and subordinate to the all-male Eight Ball Posse (see O'Malley, 1993). Another report describes that, in spite of an "equal opportunity to murder gratuitously," gangs are nevertheless "bastions of misogyny. Boys are the gatekeepers to status" (Weller, 1994). Another report describes 10 percent of Los Angeles gang members to be female: "Most often, girls join male gangs or form female offshoots of male ones. While the females may have their own leaders, most take their orders from the males" (Sikes, 1994). See for additional information Bjorkqvist and Niemela (1992); Campbell (1984); Campbell (1990); Dunham (1995); Hooks and Green (1993); and Taylor (1990).

14. Based on Table 7.1, Daly and Wilson (1988): 147, 148. Examining the most disturbing sorts of murders — serial killings and mass murders — places the male contribution into even sharper focus. See Editors (1992a, 1992b); Nash (1973). The reality of a demonic male made sense to premodern thinkers from Dante to Dostoevski.

15. Male violence exists. Feminists are more acutely aware of that fact than any other group. But in order to do anything about it, one must first understand what it is and how it manages to perpetuate itself. Why demonic males? In trying to answer that question, traditional feminist thinkers have explored several possibilities, but almost always they have located the ultimate blame for things strictly in the house of culture.

Nancy Chodorow introduced the first complete psychoanalytic theory of male temperament with *The Reproduction of Mothering* (1978). Chodorow's thinking (the dynamics of which were almost simultaneously being sketched in anthropological terms by anthropologist Sherry Ortner [see Ortner, 1974]) focuses not on the problem of male violence so much as what she regards to be a universal and universally harmful male "domination." Chodorow believes that men have become socially dominant over women as a "tenacious, almost transhistorical" fact, because men operate in the public arena whereas women find themselves corralled in the domestic. They become mothers, both producers and nurturers of children. "Women's mothering," she writes, "is one of the few universal and enduring elements in the sexual division of labor." Men live in the public, women in the private. Public defines private; and, therefore, men dominate women.

Women work in the domestic sphere, Chodorow argues, because the necessity of childbirth has been confused with the accident of child care. Biology builds women as mothers; psychodynamics constructs mothering — an activity that could be performed equally well, even interchangeably, by men and women. Women, not men, are in fact the nurturers of any social system because of what Chodorow regards as a Freudian Gordian knot: deep character formation that produces and reproduces gender. Using a feminist variation of the Oedipal theme, Chodorow argues that during the crucial, character-forming stages of development, when children are pressed warmly to their mothers' breasts and gradually learn to recognize themselves as psychosexual islands in the infinite maternal ocean ("ego formation"), little girls come to identify with the mother, that great object of first desire, whereas little boys come to disidentify. As a consequence, daughters develop the virtues of attachment: empathy, an affinity for relationships and for consensus-driven action. And sons tend to develop the virtues of detachment. They become competitive, more individualistic, and more inclined to abstract and objectify. The cycle, begun in a female mother's loving arms, approaches completion as girls grow up and find themselves psychologically prepared to be primary caregivers yet unequipped to engage in the public sphere, whereas boys grow up vice versa. Mothering has thus been reproduced.

Despite its weaknesses, Chodorow's theory demonstrated that it was possible to talk about fundamental gender difference in a provocative and positive way. She and the difference theorists who followed — including Carol Gilligan, who postulated a difference in men's and women's ethical perceptions (*In a Different Voice*, 1982); Deborah Tannen, who argued for different linguistic habits (*You Just Don't Understand*, 1990; and *That's Not What I Meant*, 1986); and Sara Ruddick, who explored issues of gender and peace (*Maternal Thinking*, 1990) — even as they sometimes discarded or moved beyond Chodorow's psychoanalysis, nonetheless created a discussion of difference that moved away from simplistic caricature toward something more complex and interesting.

Sexist stereotyping requires neither serious scholarship nor hard thinking to appreciate. Yet traditional feminists have too often seemed to regard any discussion of gender difference as old sexism in new clothes and have thus cut short the larger movement's own conversation. Katha Pollitt, writing in *The Nation*'s last issue for 1992, for example, attacks "the difference feminists" not so much because they are wrong but because their ideas are politically inexpedient. This crafty, cranky piece, "Are Women Morally Superior to Men?" takes to the woodshed Chodorow, Gilligan, Tannen, and Ruddick for their "ascription of particular virtues" to women (see Pollitt, 1992).

Starting with the observation that "the media like to caricature feminism as denying the existence of sexual differences," Pollitt caricatures the feminism of difference ("We are wiser than you poor deluded menfolk . . . so will you please-please-please listen to your moms?") before pushing it down the slippery slope of free association: "What is female? Nature. Blood. Milk. Communal gatherings. The moon. Quilts." Her own caricature of difference feminism, Pollitt continues, is just like all the sexist caricatures men have always invented to keep women in their place, like Aeschylus's celebration of law as the triumph of male order over female chaos in *The Eumenides*, or like Ayatollah Khomeini's denial of judgeships to women because they are too nice. Gender difference, so Pollitt concludes, is dangerous thinking. Though many women may find the new ideas "flattering," they should be aware that men like them because they "let men off the hook."

16. Once again, we find ourselves in the long shadow of that remarkable woman whose ideas and energy inspired generations of college students to challenge received wisdom and to fight against the assumption that mere genetic codes could constrain individual or social choices. And just as she used her experience in Samoa to support her thesis that culture was vastly more important than biology, so Mead returned again to the Pacific tropics, this time New Guinea, searching among primitive societies living in the mountains and forests for evidence about how fully culture determines sex roles or gender.

She left for New Guinea in 1931 accompanied by her second husband, anthropologist Reo Fortune, and together the two of them set out on a difficult trek across the Torricelli range into the interior of New Guinea. They stopped when their porters abandoned them at the village of a people Mead called the Mountain Arapesh, who thus became her first subject for study. In 1932 the anthropologist couple moved on to the Yuat River where they fell in with a

second tribe, the Mundugumor. And finally, in 1933, they met on
the Sepik River an English anthropologist named Gregory Bateson
who introduced them to their third subject culture, the Tchambuli.
Mead eventually divorced Fortune, married Bateson, and published
in 1935 *Sex and Temperament in Three Primitive Societies*, the full
elaboration of her theory that, although nature may produce a few
basic personality types, culture is the force that arbitrates, that assigns them to become the defining style for one sex or the other.
"Our own society," so she argued most reasonably in *Sex and Temperament*, "assigns different rôles to the two sexes, surrounds them
from birth with an expectation of different behaviour, plays out the
whole drama of courtship, marriage, and parenthood in terms of
types of behaviour believed to be innate and therefore appropriate
for one sex or for the other." But of course, she continued, our own
society confuses expectation with expression; and it is all too easy
to find one's "critical imagination handicapped by European cultural tradition" (Mead [1935; 1963]: ix and x. See also Mead [1949].)

Because Mead conveyed her conclusions about Samoa so convincingly, she has become a prime target for challenges to an extreme position of cultural determinism. And in the same way, the
very strength of her conclusions about sex roles in New Guinea
draws us to challenge Mead in this second area. Her broad outlines
and theoretical assessments of those three New Guinea societies
would in fact be repeated again and again — with little critical examination of the actual data — in popular works and in textbooks
introducing anthropology, psychology, and sociology to college students. A 1986 survey of sixty-one textbooks in psychology found
Margaret Mead to be the most commonly cited anthropologist, with
Sex and Temperament her most popular work. And from among
fifty-one college sociology textbooks, she shared the top frequency
of citations with her friend and colleague of similar ideological ori-
entation, Ruth Benedict. What generated all this acclaim was her
conclusion that nurture far more than nature creates the average
behavioral differences of the two sexes — a concept many people
find neither peculiar nor unfamiliar.

Cultural determinism led Mead to expect she would find extreme variability in the sex-roles constructed by three separate,
primitive societies. The theory led her to imagine and to promote
the idea that she had found such variability. Some contemporary and subsequent anthropological studies conducted in New

Guinea suggest Mead occasionally ignored information that would too openly contradict her thesis. (See, for instance, Fortune [1939]; Tuzin [1976; 1980]; Gerwertz [1981]; and Gerwertz and Errington [1991].) But by her own account, by the evidence of her own writing, instead of demonstrating variability all three societies actually turned out to be remarkably rigid and predictable in their expectations of gender-specific behavior. Instead of contradicting Western stereotypical sex roles, all three actually seem to reinforce them virtually to the point of parody. In all three societies, according to Mead's own report, women are responsible for all the daily cooking and domestic chores, and most or nearly all infant and child care. In all three societies, girls and women are legally regarded as the property of their fathers, brothers, and husbands. All three institutionalize a sexual double standard in favor of males — that is, only men are allowed to express sexual extroversion by having multiple spouses. In all three, only men are expected to engage in war; and, once again in all three, only men are somewhat likely or quite likely to engage in violence outside warfare. We are given evidence for all three that some men beat their wives, but no women, so far as the data inform us, beat their husbands — with a single exception from among the gentle and feminine Arapesh, where Mead describes a "wild creature" beating her husband defensively, in immediate response to his own attempt to beat her.

17. Porter (1986): 232.
18. Brownmiller (1975): 56–63.
19. Brownmiller (1975): 78–86.
20. Keegan (1993): 24, 25.
21. Keegan (1993): 28.
22. Arnhart (1990); Las Casas (1542; 1992).
23. Keegan (1993): 106–115.
24. Keegan (1993): 32–40.
25. Fisher (1992): 286.
26. Holloway (1994): 83.
27. Some 84.5 percent of the countries in a ninety-nation sample. In just about all households for 18.8 percent; at times severe enough to cause death or serious injury in 46.6 percent. Levinson (1989): 31. See also Heise, Pitanguy, and Germain (1994).
28. Holloway (1994): 77.
29. Holloway (1994): 80.
30. There is also the language problem. Emily Nasrallah's survey of

important Muslim feminists from the nineteenth and early twentieth centuries, *Women Pioneers* (1986), has not been translated out of Arabic. See Mernissi (1994): 127–130.

31. Raghavan, Shahriar, and Qureshi (1994): 36, 37.

32. Fisher (1992): 281.

33. Mead (1949): x. Her discussion of the "mistake" of the Vaërtings in imagining a matriarchy as the mirror image of patriarchy. Fisher (1992): 283. "It may be noted that I am defining matriarchy as the mirror image of patriarchy. Using that definition, I would conclude that no matriarchal society has ever existed." Lerner (1986): 31.

34. As quoted in Lerner (1986): 22.

35. Though Lerner likes to suggest a historical progression, she is the first to admit how sketchy the actual historical evidence is. "If we remember that we are here describing a historical period in which even the formal law codes have not yet been written, we can begin to appreciate how deeply rooted patriarchal gender definitions are in Western civilization." Lerner (1986): 75.

36. Lerner (1986): 18. "There is now a rich body of modern anthropological evidence available which describes relatively egalitarian societal arrangements and complex and varied solutions to the problem of divison of labor." Lerner (1986): 29.

37. Lerner (1986): 53.

38. Friedl (1975): 42.

39. To call Australian Aboriginal societies egalitarian requires some fancy footwork: "The coexistence of egalitarian and hierarchical tendencies is of course universal in human societies, but since in the Australian Aboriginal case the weighting clearly favours egalitarianism, it obliges the observer to discern when, why and for what duration inequalities of status and rights are manifested as a contrary tendency." Tonkinson (1988): 152.

40. Turnbull (1982): 153.

41. Turnbull (1965):127, 287, 271.

42. Sanday (1981): 17. Is rape a human universal? Many feminists insist it is not. Broude and Green (1976) report it absent in almost one quarter of indexed societies. Sanday (1981), examining some ninety-five societies, declares about half (forty-five) to be "rape-free" and another seventeen to be "rape-prone." But what does "rape-free" mean in these cases? Sometimes it simply means that behavior we would consider rape is not so considered in that society. Allan R.

Holmberg (1969), in his monograph on the Siriono of eastern Bo-livia, states he has "heard of no cases of rape." And yet when a man "uses a certain amount of force in seducing a potential spouse . . . that is not regarded as rape" (168, 169). As Sanday clarifies in "Rape and the Silencing of the Feminine" (1986), when she says "rape-free" she really means "relatively rape-free" (84) — as in the case of the Minangkabau of West Sumatra, whose police reports for 1981 noted "only" twenty-eight rapes in a population of 3 million. It is obvious that different societies have markedly different frequen-cies of rape; it is also obvious that different societies define very differently what constitutes rape or what constitutes a sexual viola-tion that should be reported to or handled by the police. Rape is typically underreported and must be underreported variably in vari-ous cultures, so that comparing police reports may well be mislead-ing. Nonetheless, Sanday's "low" rate for the Minangkabau (1 per 107,000) is actually a good deal higher than the rape rate recorded in England during the middle of this century (1 per 172,000 in 1947; 1 per 140,000 in 1954). Yet Sanday never lists twentieth-century Eng-land as having been "rape-free." Reported rape rates for England have increased rapidly since then, to be sure, but it is not clear how much this increase represents an increase in crime or an im-provement in crime reporting. In 1980 in England, the rape rate reached approximately 1 per 44,000, a frequency still not much more than twice that for the "rape-free" Minangkabau. Calcula-tions are based on figures reported by Sanday (1986): 84; and (for England) Tempkin (1986): 20. Craig Palmer, in "Is Rape a Cultural Universal" (1989a), has reviewed the claims of evidence on "rape-free" societies presented by Sanday (1981), also by Broude and Greene (1976), and demonstrated convincingly that those authors consistently underreported or ignored what little ethnographic data we have.

43. Turnbull (1965): 121.
44. Palmer points this contradiction out in Palmer (1989a).
45. Shostak (1981): 246.
46. Lee (1979): 454.
47. Lee (1982): 45.
48. Lee (1979): 376.
49. Lee (1982): 44. See also Harris (1989): 280.
50. Shostak (1981): 228.

51. Shostak (1981): 311.
52. Shostak (1981): 313.
53. Gowaty (1992); Hrdy (1981); Small (1991); Smuts (1992); and Smuts and Smuts (1993). See also Silk (1993).
54. See especially Smuts (1992; 1995).

7. RELATIONSHIP VIOLENCE

1. de Waal (1982).
2. de Waal (1986).
3. de Waal (1989): 65.
4. Jones, Martin, and Pilbeam (1992).
5. Jeffrey Schwartz's *The Red Ape*, for example, an extended argument that orangutans and not chimpanzees are really *Homo sapiens*'s closest living relative, fails to mention the rape evidence except in the mildest terms, a single reference to the "occasional sexual attack on a female by an overly aggressive young male." Schwartz (1987): 14.
6. Some people feel so concerned about the possibility of committing the naturalistic fallacy (the idea that what is natural is morally justifiable) that they argue against using the word "rape" for forced copulations among animals (e.g., Estep and Bruce [1981]). There are other worries, such as the fact that "rape" sounds sensational. But to us, it is important not to ignore the parallels. See also Palmer (1989).
7. Mitani's study of 179 copulations found only one initiated by the female, who placed her genitalia in front of an adult male's face. All others were initiated by males, with 88 percent being forced. A far cry from Galdikas (1981: 289), who found that female proceptive behavior (soliciting sexual advances) led to intromission in 23 out of a total of 52 observed copulations. These are different study populations from different parts of Borneo. It could be that the vastly different degrees of female proceptive behavior reflect differences of subjects, observers, or both. Mitani (1985).
8. This is from Galdikas, who found a mean of 10.8 minutes. Galdikas (1981): 287.
9. Weights from Rodman and Mitani (1987).
10. Galdikas (1981): 295.
11. Galdikas saw approaches by females to big males who called, but when John Mitani played back long calls, he found no evidence of

females approaching his tape-recorder. The attractive effect of long calls is therefore still uncertain.

12. Kingsley (1988).
13. Quote from MacKinnon (1974): 10. This quotation refers to his work in both Borneo and Sumatra, omitting a reference to Sumatran tigers.
14. MacKinnon (1971): 176.
15. Galdikas (1981). She and her colleagues witnessed 52 copulations or attempts at copulation (282). "The majority of copulations (64%)," she notes on page 284, "occur within consortship," which suggests 33 copulations in this study (64 percent of 52). Later (292) she indicates there were 15 "brief sexual contacts" in a "nonconsort context." The distinction between contexts is important partly because Galdikas found most rapes to occur outside consortship: "nonconsort copulations primarily involved subadult males (95%) with the vast majority of copulations (84%) forced" (293). Some much smaller percentage of the remaining copulations were forced as well. Elsewhere, Galdikas declares more succinctly: "Although most copulations . . . were cooperative, some involved elements of 'rape'" (287).
16. Galdikas (1981) quotations from pages 287, 288.
17. Mitani (1985): 396.
18. Rijksen (1978).
19. Galdikas (1995): 174, 175.
20. Galdikas (1995): 294.
21. Palmer (1989b).
22. Thornhill (1979): 100.
23. Other theorists take such cases into account by offering a weaker version of the fertilization-tactic hypothesis. This weaker version (which has also been applied to thinking about rape among humans) says that rape behavior has not been directly shaped by natural selection, but instead is an accidental byproduct of some other process. Rape happens regularly among ducks, according to this theory, merely because it brings together other serious components of mating strategy that do work, such as males being easily aroused, males being willing to use violence, and females being attractive to males. But whether in the strong or the weak theory, duck rape is still considered by most biologists to be simply a fertilization tactic. Donald Symons argued for this perspective in his 1979 book. For a recent review of the issues, see Palmer (1991).

24. Rijksen (1978). See page 264.
25. Smuts and Smuts (1993). See also Clutton-Brock and Parker (1995) for the theory of sexual coercion.
26. Remarkably little is known of the commonest form of stranger rape, which occurs during war. For a recent review, see Swiss and Giller (1993). See Lerner (1986, Chapter 4) for an account of the overwhelming historical evidence for the enslavement and rape of female prisoners.
27. Goodall (1986): 481.
28. Goodall (1986): 471–477.
29. Fertile females conceived in 26 percent of all consortships, compared to 16 percent of group-mating situations. See Goodall (1986).
30. Smuts and Smuts (1993). Ingrassia and Beck (1994) provide a good if brief survey of human domestic violence.
31. du Chaillu (1861): 70, 71.
32. Fossey (1983): 70.
33. Watts (1989).
34. Tiger's relationship with Nunki's group was described by Watts (1994). David Watts told us some additional details.
35. In 1995 Roger Fouts was taken by the television program 20/20 to the Laboratory for Experimental Medicine and Surgery in Primates in New York State in order to meet Booee, a chimpanzee who as a youngster had learned to communicate with Fouts using signs from American Sign Language. Subsequently, Booee had been sold by his owner (not Fouts) to the research laboratory for hepatitis (and, later, AIDS) research. Booee had had no communication with any signing individual for many years and had not even seen Fouts for seventeen years. Fouts, dressed in white lab coat and his face covered with a white protective mask, stood in front of his former student's cage and made some signs. Booee immediately responded, and surprised the human by producing a special sign, Booee's unique nickname for Fouts. Prison had not dulled the chimpanzee's memory.

8. THE PRICE OF FREEDOM

1. Kruuk (1972). See especially the movie *Eternal Enemies* (Joubert and Joubert [1992]), and the accompanying *National Geographic* article (Joubert [1994]).
2. Territorial fights normally arise when neighbors clash in boundary

areas. They can also occur when one clan leaves its own territory on a "commuting trip" to hunt distant migratory herds of wildebeest (Hofer and East [1993]). In other areas, on the other hand, clans reside permanently in "islands" of prey-rich ground, separated from their neighbors by prey-poor regions. In such places territorial encounters are so rare that no severe aggression was seen during 517 days of observation (Frank, 1986).

3. Hans Kruuk was one of the first to see it. He was in the Ngorongoro Crater in Tanzania. "September 1967, late evening. One hyena from the Mungi clan began to chase a female wildebeest, and with the help of several others pulled it down about 200 meters inside the range of the Scratching Rocks hyenas, in an area where the boundary was very well defined. More Mungi hyenas joined to share in the kill until all together there were about twenty. But at the same time, Scratching Rocks hyenas noticed that this was happening and ran up from all directions. There must have been approximately forty of them (it was difficult to count in the dark) advancing to attack the eating Mungi hyenas. The two groups mixed with an uproar of calls, but within seconds the two sides parted again and the Mungi hyenas ran away, briefly pursued by the Scratching Rocks hyenas, who then returned to the carcass. About a dozen of the Scratching Rocks hyenas, though, grabbed one of the Mungi males and bit him wherever they could — especially in the belly, the feet, and the ears. The victim was completely covered by his attackers, who proceeded to maul him for about 10 minutes while their clan fellows were eating the wildebeest. The Mungi male was literally pulled apart, and when I later studied the injuries more closely, it appeared that his ears were bitten off and so were his feet and testicles, he was paralyzed by a spinal injury, had large gashes in the hind legs and belly, and subcutaneous hemorrhages all over. . . . The next morning I found a hyena eating from the carcass and saw evidence that more had been there; about one-third of the internal organs and muscles had been eaten. Cannibals!"

4. For overviews of the importance of infanticide across different species, see Hausfater and Hrdy (1984); and Parmigiani and vom Saal (1994).

5. The debate was reopened most recently by Bartlett, Sussman, and Cheverud (1993). See the recent interchange: Sussman, Cheverud, and Bartlett (1994); with Hrdy, Janson, and van Schaik (1995).

6. The earliest record is from a 220-million-year-old dinosaur, the 6-

foot-tall *Coelophysis:* a baby inside its rib cage testified to cannibalism, and therefore probably to infanticide.

7. *Queen of Beasts* was produced by Alan Root for Survival Anglia. It was filmed in the Serengeti National Park and based on the lion studies of Anne Pusey and Craig Packer.

8. Packer and Pusey (1983).

9. The concept that infanticide is reponsible for the existence and form of social groups was proposed by Wrangham (1979), supported for gorillas by Watts (1989), applied to monogamous primates by von Schaik and Dunbar (1990), and to lemur groups by van Schaik and Kappeler (1993). Its significance for primates in general was discussed by Smuts and Smuts (1993).

10. Nishida and Kawanaka (1985).

11. L. David Mech (personal communication to RWW) reported three cases of wolves killed while their packs were trespassing, probably killed by the resident pack. In addition, one resident was killed by a trespassing pack. The trespassing packs were considered to be in search of food and encountered the local packs by chance. See Mech (1977); also Harrington (1987).

12. Kruuk (1972); East and Hofer (1991).

13. *Eternal Enemies,* filmed in the Savuti area of Chobe National Park, northern Botswana: Joubert and Joubert (1992). See also Joubert (1994).

14. Most interactions between lion prides merely involve scent-marking, roaring, and retreats. However, occasional deaths have been recorded in Serengeti, as in Chobe, including cases of prides killing intruder males and females (Schaller, 1972). Brian Bertram (1978) records the death of a male due to a bite to the spine.

15. Keegan (1993): 111–112.

16. This argument is elaborated by Manson and Wrangham (1991).

17. Bartz and Hölldobler (1982); Hölldobler and Lumsden (1980); Hölldobler (1976); and Hölldobler (1981).

18. Moorehead (1960): 46–66.

19. Golding (1954); Shakur (1993).

20. Clan sizes were 54 (mean of 8 clans in Ngorongoro: Kruuk [1972]), 52 (Frank [1986]), 43 (Cooper [1989]), and 47 (median of 7 clans in Serengeti: Hofer and East [1993]). Spotted hyena clans have also been studied in the southern Kalahari, Kruger (South Africa) and Etosha (Namibia). These sites have low prey densities and smaller clans (8, 11, and 21 respectively). See review by Hofer and East (1993a).

21. The fact that, contrary to frequent claims, female spotted hyenas are no larger than males is discussed by Frank (1986). Frank notes that females have the same linear dimensions, but are heavier than males. This difference comes from females being fatter and likely to have a fuller stomach than males. So females are not dominant because they are bigger. Rather, they are bigger as a result of being dominant.

22. Our discussion assumes that males suffer more than females from intercommunity violence, which certainly seems to be the case. Nevertheless, mothers and their infants are sometimes attacked, leading occasionally to the death of a female (see Chapter 1) or more often to the death of her infant (Goodall [1986]: 522). Too little is known to account for these attacks, but most authors think they are part of a system for recruiting additional reproductive females. Intrigued by the killing of Madam Bee, Wolf and Schulman (1984) suggested that males might be particularly brutal toward females who are too old to reproduce much longer. Goodall (1986: 524–525) proposed that the killing of Madam Bee fit a strategy of the killer males' trying to recruit adolescents by destroying the mother-daughter bond. The infanticidal attacks can also be interpreted as a case of relationship violence in the same manner as gorilla infanticide. By this view, a mother whose infant is killed learns that the males of her current community are inadequate as her defenders, so that a strategically sensible move from her point of view would be to move into the killers' community, where her next infant would presumably be safer (Wrangham, 1979). As Goodall (1986) concluded: "More facts are badly needed."

23. Nishida and Kawanaka (1985). For a broad account of chimpanzee life in the Mahale Mountains, see Nishida (1990).

24. Stanford (1995).

25. Manson and Wrangham (1991).

26. Wrangham and others (1992).

27. Chapman, Wrangham, and Chapman (1995).

28. Janson and Goldsmith (1995). See also Wrangham, Gittleman, and Chapman (1993). The principle of groups being as large as possible applies well to frugivorous primates, but not to the leaf eaters.

29. Baker and Smuts (1994).

30. van Hooff and van Schaik (1994).

31. Keegan (1993): 124.

32. Keegan (1993): 124.

33. Archaeological evidence provides a real if sketchy look at the primitive past. Evidence includes Trinkaus (1978), who describes the signs of severe wounds from apparent violent encounters — such as the flattening of the outside area of an eye socket — in skeletons of Neanderthals who died 70,000 years ago. See also Anderson (1968); Dastugue and de Lumley (1976); Klein (1989): 333–334; and Wendorf and Schild (1986). Recent excavations in a Nubian graveyard of modern Sudan turned up 58 skeletons of men, women, and children buried somewhere around 10,000 to 12,000 B.C. Nearly half the skeletons showed clear evidence of violent death: Wendorf (1986). Is the modern period an improvement? In *Statistics of Deadly Quarrels*, Lewis Richardson (1960) presents the calculation that 59 million people were killed by their fellow humans during the century and a quarter between 1820 and 1945. See also Freeman (1964) and Keeley (1996).

9. LEGACIES

1. Strier (1992).
2. Muriquis *(Brachyteles arachnoides)* have been studied most extensively by Karen Strier (Strier [1992]). Why male muriquis are so unaggressive is still not clear. Part of the answer is that they compete through sperm competition. But there are other species in which males also compete by massive overproduction of sperm, yet nevertheless are aggressive also, such as chimpanzees. Their unusual style of locomotion may be responsible for preventing males from getting larger than females, and thereby for females being unafraid of males. Females choose which males to mate, and would presumably choose not to mate with a male who was aggressive.
3. Strier (1992): 4.
4. As mentioned in the footnote in Chapter 6, our use of temperament was inspired by Clarke and Boinski (1995); but we have expanded the conception beyond their original, very specific meaning.
5. Canine tooth development is rather closely linked to the intensity of male-male aggression in primates, and probably in many mammals. See Plavcan and van Schaik (1992) and Plavcan and van Schaik (1994). Selection for aggression also affects female canine length (Harvey, Kavanagh, and Clutton-Brock [1978]). The idea that human canines are small because hominids developed an alternative style

of fighting was proposed by Darwin and many others since. Our discussion of this classic proposal differs from most previous ones by imagining that the shift to fistfighting could have occurred even without significant use of hand-held weapons. Incidentally, it is still unclear why selection rather quickly caused canines to be shortened once they were not needed for fighting. One possibility is that the canine was pressed into service as an additional incisor, i.e., that it was needed as a food-cutting tooth. See Greenfield (1992) and Plavcan and others (1995), for a review.

6. Jarman (1989).
7. Tanner (1978); Malina and Bouchard (1991); Bribiescas (1996); Jamison (1978).
8. Wolpoff (1980): 178.
9. Howard (1983b): 34.
10. Howard (1983a): 22.
11. First-time mothers tend to have singletons, however. See Frank, Glickman, and Licht (1991).
12. Frank, Glickman, and Licht (1991). The accounts of fighting are based on five litters born at the University of California, Berkeley.
13. Frank, Glickman, and Licht (1991): 704.
14. Accepted wisdom: see Frank, Weidele, and Glickman (1995); Sapolsky (1994). Advantage to being high-ranking: Frank, Weidele, and Glickman (1995). Masculinizing effects of a mother's high androgen: Ward (1978): 4. For a review of nineteen studies of the effects of an abnormal hormonal environment on the fetus during pregnancy: Reinisch, Ziemba-Davis, and Sanders (1991). Clinical pseudohermaphrodism: Ehrhardt and Baker (1978). Risks of maternal mortality and stillbirths among hyenas: Frank, Weidele, and Glickman (1995); and Glickman and others (1992).
15. Goodall (1986): 410.
16. Goodall (1986): 426.
17. Damasio (1994).
18. Baker and Smuts (1994).
19. Interestingly, a few individuals buck the trend. Jomeo, in Gombe, was famous for his failure to challenge for status, despite being one of the largest males in the community. Slim, in Kibale, is an equivalent of Jomeo. The relative importance of experience and genes in shaping these unusually timid personalities is unknown.
20. Howard (1983a): 10.
21. The history of the Peloponnesian War is described by Finley (1963).

22. Howard (1983a): 15.
23. Shakur (1993): 56.
24. Arnhart (1995): 492, 293.
25. Wolff (1995).
26. Sherif and others (1961).
27. Blake and Mouton (1962); Mouton (1979). See also Rabbie (1992).
28. Ethnocentrism was defined as "the view of things in which one's own group is the center of everything. . . . Each group nourishes its own pride and vanity, boasts itself superior . . . and looks with contempt on outsiders" (Sumner [1906]: 12). Ethnocentrism was found worldwide in a survey of twenty groups by Brewer (1979).
29. Stephan and Stephan (1990): 431.
30. Brown (1986).
31. Las Casas (1542; 1992): 29.
32. Brown (1970).
33. Darwin (1871).
34. For elaboration of the relationship between intergroup fighting and morality, see Alexander (1987) and Arnhart (1995).
35. Raper (1933): 143–144.
36. Buford (1992): 204, 205.
37. Simmel (1950).
38. Baron and Byrne (1977): 586.

10. THE GENTLE APE

1. Kano (1990).
2. In 1984 a government capture operation resulted in the loss of ten to twenty bonobos. With the economic hard times of the 1990s poaching in the Wamba area increased (Thompson-Handler, Malenky and Reinharz [1995]: 29). In February 1994, Kano was shown the hand of a bonobo that had apparently been killed for meat. Increasing mobility combined with the stagnant economy of Zaïre will continue to place new threats on bonobos everywhere.
3. Schwarz (1929). Schwarz called bonobos a subspecies of chimpanzees, *Pan satyrus paniscus*. In 1933 Harold Coolidge raised the status to its presently accepted position, *Pan paniscus*, a sister species to chimpanzees. See van den Audenaerde (1984).
4. A bonobo skull had lain in the British Museum since 1881, and

another in the Tervuren Museum since 1910: van den Audenaerde (1984).

5. Bonobos were first called pygmy chimpanzees. For some people, *pygmy chimpanzee* is an endearing term which should be used because it has priority. But it has several disadvantages. Apart from the fact that it gives an exaggerated idea of how small they are, it pushes people toward using the term *common chimpanzee* for *Pan troglodytes*. This undermines conservation efforts. Unfortunately no alternative to *pygmy chimpanzee* is perfect, but *bonobo* — suggested by Tratz and Heck (1954) — has become popular. Tratz and Heck claimed it to be an indigenous name; probably not. Adriaan Kortlandt suggests that it was a misspelling of the town name Bolobo on the crate that brought their poor ape from the wild to Germany; Kortlandt (1993). Kano (1992) records "*elya*, plural *bilya*" as an indigenous (Mongo) name for this species throughout much of its range (42).

6. The Yerkes Regional Primate Research Center, Atlanta, Georgia.

7. Coolidge (1984).

8. Yerkes (1925): Chapter 13.

9. The view of bonobos as a species that has evolved away from a chimpanzee-like ancestor is discussed by Wrangham, McGrew, and de Waal (1994). An alternative proposal sees bonobos as more similar than chimpanzees are to their common ancestor with humans. This other view has been based mainly on similarities in the limb proportions of bonobos and australopithecines: Zihlman and others (1978). However, the strong similarity of chimpanzee and gorilla growth patterns, together with the fact that many bonobo features are a juvenilized version of the chimpanzee's, seems to us impressive evidence for chimpanzees' being the more conservative of the two species.

10. Kano's encounter was on October 28, 1973, extraordinarily late for humans to begin uncovering the lives of so close a relative: Kano (1979): 130. For a detailed description of vocal differences between chimpanzees and bonobos see de Waal (1988): 203. The high hoots of bonobos reach a peak frequency of 2.3 kHz in the lowest harmonic, compared to about 1 kHz for chimpanzees (Clark-Arcadi submitted). The average duration of bonobo hoots is 0.3 second (range up to 0.7 second), compared to an average of about 0.8 for the chimpanzee equivalent (the climax element in the pant hoot).

11. Kano (1992).
12. Kano (1992): 183–184.
13. Furuichi and Ihobe (1994). Lifelong association between mother and son is very rare among animals. Pilot whales and orcas (killer whales) are the only other species in which lifelong mother-son bonds appear to be so important.
14. Kano (1992): 189.
15. Kano (1992): 185 (Table 23).
16. Hohmann and Fruth (1993).
17. Parish (1993).
18. Idani (1991b). See also Furuichi (1989); and Kano (1992).
19. This process looks much like one of the ways that a male baboon integrates himself into a new troop, by targeting a female, showing her respect, waiting for her to give signals that he may approach, and developing a special friendship with her. Smuts (1985).
20. The inference that *hoka-hoka* is orgasmic is bolstered by experimental evidence from monkeys. Goldfoot and others (1980) showed that when stumptail monkeys are stimulated sexually, they reach a behavioral climax (similar to that seen in female bonobos) which is accompanied by physiological climax similar to the pattern in women (uterine contractions, elevated blood flow). Behavioral and physiological climax occurs in women at the time that they experience orgasm. Bonobos and other primates that show these climaxes may also have orgasmic feelings. They certainly give every appearance of enjoying sex enormously.
21. Kano (1992): 176 (Table 23). Out of 259 aggressive interactions among adults, in which both sexes were present in approximately equal numbers, 3.5 percent were among females, compared to 62.9 percent among males. The remainder were aggression by males toward females (31.7 percent) or by females toward males (1.9 percent). The frequency of interaction was abnormally high because encounters were at a feeding station.
22. In a comparable way, another primate (red colobus) has been found where female power has reduced the male-male bond. Starin (1994).
23. Furuichi and Ihobe (1994).
24. Furuichi and Ihobe (1994). Proximity indices were calculated as the time spent within 3 meters divided by the total time each member of a dyad was observed. The average proximity index across dyads was .03–.04 for both species.
25. Furuichi and Ihobe (1994). Grooming indices were the number of

observations in which the focal male groomed with a male, divided by the total time each member of a dyad was observed. The average grooming index was about .01 for both species.

26. Goodall (1991).

27. Furuichi and Ihobe (1994).

28. Furuichi and Ihobe (1994), Muroyama and Sugiyama (1994).

29. Nishida and others (1991).

30. Furuichi and Ihobe (1994): 220: For example, two out of ninety agonistic interactions were over copulations, though copulations occur daily. Comparison with chimpanzees is difficult because chimpanzees copulate on fewer days.

31. Like chimpanzees, female bonobos have sexual swellings that become large and highly visible in response to female sex hormones. Males can discriminate females with swellings from those without; and because they mate females more often when the swelling is firmer, it appears that bonobo males can tell which part of her sexual cycle the female is in. In chimpanzees, the sexual swelling disappears within a day of ovulation, shortly after male-male competition for mating reaches a peak. In bonobos, the sexual swelling is maintained, and there is no evidence of male interest peaking on any particular day. Furuichi (1987). See also Furuichi (1992).

32. Wrangham (1993).

33. *Humanae vitae*, as cited by Wickler (1967).

34. de Waal (1990).

35. Males are less aggressive than average (among mammals) while females are more sexual. We find a similar combination in other species. Most striking is the muriqui, which among South American monkeys is behaviorally the nearest thing to a bonobo. Like bonobos, male muriquis live with their male kin and defend a range. Like bonobos, female muriquis are codominant with males, and they mate frequently. And like bonobos, males do not fight for matings; instead, they tolerate other males mating even in full view. Muriquis provide evidence supporting the idea that female power favors relaxed, public, nonconceptive sex. Strier (1994).

36. Idani (1991a).

37. E-group was one of the first communities watched by Kano and his team. It later split into two independent communities, E1 and E2. The friendly intercommunity interactions described here involved E1-group, abbreviated here to E-group.

38. The Lomako field study, initiated by Noel and Alison Badrian, con-

tinued by a Stony Brook team, and now run by Barbara Fruth and Gotfried Hohmann, has produced results similar to those at Wamba. See, for example, White (1992).

39. Descriptions are given by Boesch and Boesch (1989). Explanations for cooperative hunting are discussed by Boesch (1994).

40. Wrangham and Riss (1990). Stanford and others (1994).

41. For Gombe and Taï, see Stanford and Boesch articles. For Mahale, see Uehara and others (1992). The Kibale data have not been published in detail. Red colobus and guerezas are both killed frequently by chimpanzees.

42. Badrian and Malenky (1984). See also Hohmann and Fruth (1993), who report a total of 5 kills, the largest estimated at 8–10 kilograms. The antelope were young duikers *Cephalophus nigrifrons* and *C. dorsalis.*

43. Flying squirrels *Uromastyx* are recorded as prey at Wamba; see Kano (1992): 106; see also Ihobe (1992); for Lilungu, Sabater Pi and others (1993). Lilungu bonobos were also recorded eating a fruit bat.

44. Sabater Pi and others (1993).

45. Robert C. Bailey (personal communication). See Bailey (1991).

46. Ihobe (1990).

47. In Kibale, red colobus often go out of their way to approach and attack chimpanzees. I once watched three adult males chase eight male chimpanzees out of a tree crown.

48. There are no red colobus at Lomako, but other monkeys are common, including species eaten elsewhere by chimpanzees. See Hohmann and Fruth (1993).

49. Dart (1953).

50. Lorenz (1963; 1966). See Huntingford (1976) for a balanced view, noting that Lorenz's distinction between predation and aggression may not be as clear as he argued.

11. MESSAGE FROM THE SOUTHERN FORESTS

1. Yerkes (1925): 246–248.

2. Kuroda (1979); and Kano (1982).

3. Wrangham (1986).

4. Chapman, White, and Wrangham (1994).

5. Badrian and Badrian (1984).

6. Kano (1979).
7. Malenky and Wrangham (1994); and Malenky and others (1994).
8. Chapman, White, and Wrangham (1994).
9. Wrangham and others (1996).
10. Wrangham (1986).
11. Kinzey (1984).
12. Colyn and others (1991).
13. In the west, there are mountain ranges in Cameroon and Gabon; in the east, in Uganda, Rwanda, and Zaïre. Some of the modern-day mountains are recent, but even as far back as 3 million years ago there were mountainous areas in these regions that should have contained refuges for gorillas.
14. Wolfheim (1983): 686.
15. Sugiyama (1988).
16. Baker and Smuts (1994).
17. Goodall (1990): 76–80. Infanticide by females might serve the killers by inducing rival mothers to feed elsewhere, reducing competition for food.
18. This is seen occasionally in the wild. For example, Goodall (1990): 166–167. It has been seen most prominently, however, in stable groups in captivity. See de Waal (1982); and Baker and Smuts (1994).
19. The technical meaning of social bonds varies a little between studies. We mean a bond to be evidenced by a statistically significant matrix correlation between two independent types of affiliative interaction. For example, if the frequency of female-female grooming (in relation to the amount of time spent together) is correlated with the frequency of female-female coalitionary support within a given set of individuals, we take this as evidence of bonding. See Smuts (1985) for an elaboration of principles; and for a specific example showing male-male bonds in chimpanzees, see Wrangham, Clark, and Isabirye-Basuta (1992).
20. de Waal (1982). See also Baker and Smuts (1994).
21. Chapman, White, and Wrangham (1994).
22. Chapman and White (1994).
23. Furuichi (1989) showed that female bonobos tend to be in the middle of parties. Data on sex differences in spatial arrangements of chimpanzees appear not to have been collected, but it is clear from my own observations that females are normally peripheral in parties in Gombe and Kibale.

24. The gorilla-food hypothesis for explaining the stable parties of bonobos is still being tested by data on bonobos, chimpanzees, and gorillas. Until the costs of grouping have been directly measured, we will not know whether any other factors are needed to explain the difference between bonobo and chimpanzee parties. It is possible that differences in the way the two species travel (with bonobos being more arboreal than chimpanzees) has more important effects than we have implied here. See Wrangham and others (1996).

25. The idea that the evolution of australopithecines into early humans was driven by climatic change was proposed by Vrba (1988). The climatic data are still too crude to test it adequately. It is supported by high extinction rates of forest-adapted antelopes and rodents, and an increase in the number of savanna-adapted species, around 2.4–2.5 million years ago. See Stanley (1992).

26. The Milankovitch cycles that describe changes in the earth's orbit also include changes in the tilt of the axis of rotation and in the season of the year during which the earth is closest to the sun. See Terborgh (1992) and Jones, Martin, and Pilbeam (1992).

12. TAMING THE DEMON

1. Manson and Wrangham (1991).
2. Betzig (1992) describes Roman polygyny. The details of the Chinese harem system come from van Gulik (1974). See also Dickemann (1979) for the effects of extreme power on marriage arrangements. For a historical review of polygyny within the Jewish tradition, see Rubenstein (1974). See Brodie (1946) for a description of one comparatively modern attempt to formalize polygyny for powerful males; see Fotheringham (1991) for a reaction to one informal attempt at polygyny on a massive scale; see also "Ricki Lake" (1995). Prostitution, of course, represents another expression of the polygynous inclination within (and sometimes without) officially monogamous societies since prostitution invariably supplies a male (both heterosexual and homosexual) demand. See, for example, Bullough and Bullough (1964); Hornblower (1993); and Perkins and Bennett (1985). Pornography reflects the same pattern (Stauffer and Frost [1976]).
3. Betzig (1991).

4. Lerner (1986).

5. The novel was written in 1915, serialized in Gilman's feminist monthly, *The Forerunner,* finally recovered, reappreciated, and published separately in 1979.

6. "We were now well used to seeing women not as females but as people, people of all sorts, doing every kind of work." Gilman (1915; 1979): 136.

7. Possibly the best comparisons between likely social-political systems favored by each sex can be gained from studying single-sex institutions. The comparisons will invariably be imperfect because both all men's and all women's institutions tend to operate within a larger patriarchal sphere of power. Catholic convents, for example, even as they exclude men from their daily affairs, are organized under the authority of priests and ultimately the pope. Nonetheless, studies of single-sex institutions provide some fascinating comparisons. Thomas W. Foster argues that women in prisons create "make-believe families" while men in prison create hierarchies based on power. See Foster (1979); also Giallombardo (1966); Klare (1979); Kruttschnitt (1981); Wilson (1980); and Wilson (1986).

8. The most important evidence here is that castration routinely reduces aggression in males. See Wingfield and others (1994).

9. Mitchell (1936): 663.

10. Mitchell (1936): 665.

11. As quoted in Kuntz (1995).

12. In a speech of November 6, 1938.

13. See Putnam (1993): 125, 146.

14. Berreman (1993): 366.

15. Greater social equality of Indian tribal life: Berreman (1993): 386; Chipko: Berreman (1993): 378–382.

16. Based on Beck (1995); Gleick (1995); Gibbs (1995); Leland and others (1995); Morganthau and others (1995); Weiss (1995).

17. As quoted by Shabad (1995).

18. As quoted in "Ready" (1995).

19. Sorokin (1962: 295–341) presents death rates from international war, averaged by century, for Greece (fifth to second century B.C.), Rome (fourth century B.C. to the third century A.D.) and for Europe in the seventeenth and eighteenth centuries. Small and Singer (1983: 118 and 252) present annual death rates from wars for 176 states between 1816 and 1980.

13. KAKAMA'S DOLL

1. Hayes (1951): 80–85.
2. It is possible, though difficult, to infer mental processes. Heyes (1995) gives a good sense of how carefully experiments and observations must be designed.
3. "Brain size" here implies "relative to body size." The relationship between brain and body size is complicated, so that it is still unclear to what extent the fossil record supports the notion of a steady increase in relative brain size. On the whole, most animal species appear to keep the same relative brain size for long periods. See Deacon (1990).
4. Savage-Rumbaugh and Lewin (1994).

BIBLIOGRAPHY

Adams, David B. 1983. "Why There Are So Few Women Warriors." *Behavior Science Research* 18 (3): 196–212.

Adler, Freda. 1975. *Sisters in Crime: The Rise of the New Female Criminal.* New York: McGraw-Hill.

Aiello, Leslie C., and Peter Wheeler. 1995. "The Expensive-tissue Hypothesis: The Brain and the Digestive System in Human and Primate Evolution." *Current Anthropology* 36: 199–221.

Albert, Bruce. 1989. "Yanomami 'Violence': Inclusive Fitness or Ethnographer's Representation?" *Current Anthropology* 30: 637–640.

———. 1990. "On Yanomami Warfare: A Rejoinder." *Current Anthropology* 31: 558–563.

Alexander, Richard D. 1987. *The Biology of Moral Systems.* Hawthorne, N.Y.: Aldine de Gruyter.

Amrane, Djamila. 1982. "Algeria: Anticolonial War." Translated by Richard Stites. In *Female Soldiers — Combatants or Noncombatants? Historical and Contemporary Perspectives.* Edited by Nancy Loring Goldman. Westport, Conn.: Greenwood Press: 123–135.

Anderson, J. E. 1968. "Late Paleolithic Skeletal Remains from Nubia." In *The Prehistory of Nubia*, vol. 2. Edited by Fred Wendorf. Dallas: Southern Methodist University Press: 996–1040.

Archer, John. 1988. *The Behavioral Biology of Aggression.* Cambridge: Cambridge University Press.

———. 1994. *Male Violence*. London: Routledge.

Ardrey, Robert. 1966. *The Territorial Imperative*. New York: Atheneum.

Arnhart, Larry. 1990. "Aristotle, Chimpanzees and Other Political Animals." *Biology and Social Life* 29 (3): 477–557.

———. 1995. "The New Darwinian Naturalism in Political Theory." *American Political Science Review* 89: 389–400.

Audenaerde, Thys van den. 1984. "The Tervuren Museum and the Pygmy Chimpanzee." In *The Pygmy Chimpanzee: Evolutionary Biology and Behavior.* Edited by Randall Susman. New York: Plenum Press: 3–12.

Badrian, Alison, and Noel Badrian. 1984. "Social Organization of *Pan paniscus* in the Lomako Forest, Zaïre." In *The Pygmy Chimpanzee: Evolutionary Biology and Behavior.* Edited by R. L. Susman. New York: Plenum Press: 325–346.

Badrian, Noel, and Richard Malenky. 1984. "Feeding Ecology of *Pan paniscus* in the Lomako Forest, Zaïre." In *The Pygmy Chimpanzee: Evolutionary Biology and Behavior.* Edited by R. L. Susman. New York: Plenum Press: 275–299.

Bahuchet, Serge. 1990. "The Aka Pygmies: Hunting and Gathering in the Lobaye Forest." In *Food and Nutrition in the African Rain Forest.* Edited by C. Marcel Hladik, Serge Bahuchet, and Igor de Garine. Paris: UNESCO: 19–23.

Bailey, Robert C. 1991. *The Behavioral Ecology of Efé Pygmy Men in the Ituri Forest, Zaïre.* Ann Arbor: University of Michigan Press.

Bailey, Wendy J., and others. 1992. "Reexamination of the African Hominoid Trichotomy with Additional Sequences from the Primate β-Globin Gene Cluster." *Molecular Phyogenetics and Evolution* 1: 97–135.

Baker, Kate C., and Barbara B. Smuts. 1994. "Social Relationships of Female Chimpanzees: Diversity Between Captive Social Groups." In *Chimpanzee Cultures.* Edited by Richard W. Wrangham and others. Cambridge, Mass.: Harvard University Press: 227–242.

Barlow, Nora, ed. 1958. *The Autobiography of Charles Darwin, 1809–1882.* New York: Harcourt, Brace.

Baron, Robert A., and Donn Bryne. 1977. *Social Psychology: Understanding Human Interaction.* Boston: Allyn and Bacon.

Bartlett, Thad Q., Robert W. Sussman, and James M. Cheverud. 1993. "Infant Killing in Primates: A Review of Observed Cases with Specific Reference to the Sexual Selection Hypothesis." *American Anthropologist* 95: 958–990.

Bartz, Stephan R., and Bert Hölldobler. 1982. "Colony Founding in *Myrmecocystus mimicus* Wheeler (Hymenoptera: Formicidae) and the Evolution of Foundress Associations." *Behavioral Ecology and Sociobiology* 10: 137–147.

Beck, Melinda. 1995. "Get Me Out of Here." *Newsweek* (May 1): 40–47.

Belenky, Mary Field, and others. 1986. *Women's Ways of Knowing: The Development of Self, Voice, and Mind.* New York: Basic Books.

Berreman, David D. 1993. "Sanskritization as Female Oppression in India." In *Sex and Gender Hierarchies*. Edited by Barbara Diane Miller. Cambridge: Cambridge University Press: 366–392.

Bertram, Brian. 1978. *Pride of Lions.* London: John Dent and Sons.

Betzig, Laura. 1992. "Roman Polygyny." *Ethology and Sociobiology* 13: 309–349.

Bigelow, Robert. 1969. *The Dawn Warriors: Man's Evolution Towards Peace.* Boston: Little, Brown.

Bjorkqvist, Kaj, and Pirkko Niemela. 1992. *Of Mice and Women: Aspects of Female Aggression.* Boston: Academic Press.

Blake, R. R., and J. S. Mouton. 1962. "The Intergroup Dynamics of Win-Lose Conflict and Problem-solving Collaboration in Union-Management Relations." In *Intergroup Relations and Leadership: Approaches and Research in Industrial, Ethnic, Cultural, and Political Areas.* Edited by Muzafer Sherif. New York: Wiley.

———. 1979. "Intergroup Problem Solving in Organizations: From Theory to Practice." In *The Social Psychology of Intergroup Relations.* Edited by W. G. Austin and S. Worchel. Monterey, Calif.: Brooks/Cole: 19–32.

Bloom, Anne R. 1982. "Israel: The Longest War." In *Female Soldiers — Combatants or Noncombatants? Historical and Contemporary Perspectives.* Edited by Nancy Loring Goldman. Westport, Conn.: Greenwood Press: 137–162.

Blumenschine, Robert J., and John A. Cavallo. 1992. "Scavenging and Human Evolution." *Scientific American* (October): 90–96.

Boas, Franz. 1924. "The Question of Racial Purity." *The American Mercury* 3: 163–169.

Boehm, Christopher. 1992. "Segmentary 'Warfare' and the Management of Conflict: Comparison of East African Chimpanzees and Patrilineal-patrilocal Humans." In *Coalitions and Alliances in Humans and Other Animals.* Edited by A. H. Harcourt and Frans B. M. de Waal. Oxford: Oxford University Press: 137–173.

Boesch, Christophe. 1994. "Cooperative Hunting in Wild Chimpanzees." *Animal Behaviour* 48: 653–667.

Boesch, Christophe, and Hedwige Boesch. 1989. "Hunting Behavior of Wild Chimpanzees in the Taï National Park." *American Journal of Physical Anthropology* 78: 547–573.

Bowlby, John. 1990. *Charles Darwin: A New Life.* New York: W. W. Norton.

Brain, Charles K. 1988. "New Information from the Swartkrans Cave of Relevance to 'Robust' Australopithecines." In *Evolutionary History of the "Robust" Australopithecines.* Edited by F. E. Grine. Hawthorne, N.Y.: Aldine de Gruyter: 311–324.

Brewer, M. B. 1979. "The Role of Ethnocentrism in Intergroup Conflict." In *The Social Psychology of Intergroup Relations.* Edited by W. G. Austin and S. Worchel. Monterey, Calif.: Brooks/Cole: 71–84.

Brewer, Stella. 1978. *The Forest Dwellers.* London: Collins.

Bribiescas, Richard G. 1996. "Salivary Testosterone Levels Among Aché Hunter-Gatherer Men and a Functional Interpretation of Population Variation in Testosterone Among Adult Males." *Human Nature* (in press).

Brodie, Fawn. 1946. *No Man Knows My History: The Life of Joseph Smith, the Mormon Prophet.* New York: Alfred A. Knopf.

Brown, Dee. 1970. *Bury My Heart at Wounded Knee: An Indian History of the American West.* New York: Henry Holt and Company.

Brown, Roger. 1986. *Social Psychology,* 2nd ed. New York: The Free Press.

Brownmiller, Susan. 1975. *Against Our Will: Men, Women and Rape.* New York: Fawcett Columbine.

Broude, G., and S. Greene. 1976. "Cross-Cultural Codes on Twenty Sexual Attitudes and Practices." *Ethnology* 15: 409–429.

Brunet, Michel, and others. 1995. "The First Australopithecine 2,500 Kilometres West of the Rift Valley, Chad." *Nature* 378: 273–275.

Buford, Bill. 1992. *Among the Thugs.* New York: W. W. Norton.

Bullough, Vern L., and Bonnie L. Bullough. 1964. *The History of Prostitution.* New Hyde Park: University Books.

Burkhart, Kathryn Watterson. 1973. *Women in Prison.* Garden City, N.Y.: Doubleday.

"Burundi: The Terror Behind the Putsch." 1993. *Africa Confidential* 34 (December 17): 6, 7.

Caccone, Adalgisa, and Jeffrey R. Powell. 1989. "DNA Divergence Among Hominoids." *Evolution* 43: 925–942.

Cachin, Françoise. 1989; 1992. *Gauguin: The Quest for Paradise.* New York: Harry N. Abrams.

Campbell, Anne. 1984. *The Girls in the Gang.* New York: Basil Blackwell.

———. 1990. "Female Participation in Gangs." In *Gangs in America.* Edited by C. Ronald Huff. Newbury Park, Calif.: Sage Publications.

Cartmill, Matt. 1994. *A View to a Death in the Morning: Hunting and Nature Through History.* Cambridge, Mass.: Harvard University Press.

Chagnon, Napoleon A. 1988. "Life Histories, Blood Revenge, and Warfare in a Tribal Population." *Science* 239 (February 20): 985–992.

———. 1990. "On Yanomamö Violence: Reply to Albert." *Current Anthropology* 31: 49–53.

———. 1992. *Yanomamö: The Last Days of Eden.* New York: Harcourt Brace Jovanovich.

Chapman, Colin A., and Frances J. White. 1994. "Nearest Neighbor Distances in Chimpanzees and Bonobos." *Folia Primatologica* 63: 181–191.

Chapman, Colin A., Frances J. White, and Richard W. Wrangham. 1994. "Party Size in Chimpanzees and Bonobos: A Reevaluation of Theory Based on Two Similarly Forested Sites." In *Chimpanzee Cultures.* Edited by Richard W. Wrangham and others. Cambridge, Mass.: Harvard University Press: 41–58.

Chapman, Colin A., Richard W. Wrangham, and Lauren J. Chapman. 1995. "Ecological Constraints on Group Size: An Analysis of Spider Monkey and Chimpanzee Subgroups." *Behavioral Ecology and Sociobiology* 36: 59–70.

Chodorow, Nancy. 1978. *The Reproduction of Mothering: Psychoanalysis and the Sociology of Gender.* Berkeley: University of California Press.

Clarke, A. Susan, and Sue Boinski. 1995. "Temperament in Nonhuman Primates." *American Journal of Primatology* 37: 103–125.

Clutton-Brock, Timothy H., James C. Deutsch, and R. J. C. Nefdt. 1993. "The Evolution of Ungulate Leks." *Animal Behaviour* 46: 1121–1138.

Clutton-Brock, Timothy H., and Geoffrey A. Parker. 1995. "Sexual Coercion in Animal Societies." *Animal Behaviour* 49: 1345–1365.

Colson, Elizabeth. 1993. "A Note on the Discussions at Mijas." In *Sex and Gender Hierarchies.* Edited by Barbara Diane Miller. Cambridge: Cambridge University Press: xv–xix.

Colyn, Marc, and others. 1991. "A Re-appraisal of Palaeoenvironmental History in Central Africa: Evidence for a Major Fluvial Refuge in the Zaïre Basin." *Journal of Biogeography* 18: 403–407.

"'Comfort Women' Reparations Urged." 1994. *Boston Globe* (November 22): 14.

Coolidge, Harold J. 1984. "Historical Remarks Bearing on the Discovery of *Pan paniscus.*" In *The Pygmy Chimpanzee: Evolutionary Biology and Behavior.* New York: Plenum Press: ix–xiii.

Cooper, S. M. 1989. "Clan Sizes of Spotted Hyaenas in the Savuti Region of the Chobe National Park, Botswana." *Botswana Notes and Records* 21: 121–133.

Coppens, Yves. 1994. "East Side Story: The Origin of Humankind." *Scientific American* (May): 88–95.

Crespi, Bernard J. 1986. "Size Assessment and Alternative Fighting Tactics in *Elaphrothrips tuberculatus* (Insecta: Thysanoptera)." *Animal Behaviour* 34: 1324–35.

Daly, Martin, and Margo Wilson. 1988. *Homicide.* New York: Aldine de Gruyter.

Damasio, Antonio R. 1994. *Descartes' Error: Emotion, Reason, and the Human Brain.* New York: Grosset/Putnam.

Dart, Raymond A. 1953. "The Predatory Transition from Ape to Man." *International Anthropological and Linguistic Review* 1 (4): 201–218.

Darwin, Charles. 1839; 1972. *The Voyage of the Beagle.* New York: New American Library.

———. 1871. *The Descent of Man and Selection in Relation to Sex,* vol. 1. London: John Murray.

Dastugue, J., and M.-A. de Lumley. 1976. "Les Maladies des Hommes Préhistoriques du Paleolithique et du Mésolithique." In *La Préhistoire Française,* vol 1. Edited by H. de Lumley. Paris: Centre National de la Recherche Scientifique: 612–622.

Deacon, Terrence W. 1990. "Rethinking Mammalian Brain Evolution." *American Zoologist* 30: 629–705.

de Waal, Frans B. M. 1982. *Chimpanzee Politics: Power and Sex Among the Apes.* New York: Harper and Row.

———. 1986. "The Brutal Elimination of a Rival Among Captive Male Chimpanzees." *Ethology and Sociobiology* 7: 237–251.

———. 1988. "The Communicative Repertoire of Captive Bonobos *(Pan paniscus)*, Compared to that of Chimpanzees." *Behaviour* 106: 183–251.

———. 1989. *Peacemaking Among Primates.* Cambridge, Mass.: Harvard University Press.

———. 1990. "Sociosexual Behavior Used for Tension Regulation in All Age and Sex Combinations Among Bonobos." In *Pedophilia: Biosocial Dimensions.* Edited by J. R. Feierman. New York: Springer: 378–393.

Dennett, Daniel C. 1995. *Darwin's Dangerous Idea: Evolution and the Meaning of Life.* New York: Simon and Schuster.

Dickemann, Mildred. 1979. "The Ecology of Mating Systems in Hypergynous Dowry Societies." *Social Science Information* 18: 163–195.

Djian, P., and Howard Green. 1989. "Vectorial Expansion of the Involucrin Gene and the Relatedness of Hominoids." *Proceedings of the National Academy of Sciences* 86: 8447–8451.

Doggett, Maeve E. 1993. *Marriage, Wife-Beating, and the Law in Victorian England.* Columbia, S.C.: University of South Carolina Press.

du Chaillu, Paul B. 1861. *Explorations and Adventures in Equatorial Africa: With Accounts of the Manners and Customs of the People and of the Chace of the Gorilla, Crocodile, Leopard, Elephant, Hippopotamus, and Other Animals.* London: John Murray.

Dunham, Elizabeth. 1995. "Bad Girls." *'Teen* (August): 52 ff.

Dworkin, Andrea. 1987. *Intercourse.* New York: The Free Press.

East, Marion L., and Heribert Hofer. 1991. "Loud-calling in a Female Dominated Mammalian Society. II. Behavioral Contexts and Functions of Whooping of Spotted Hyaenas, *Crocuta crocuta.*" *Animal Behaviour* 42: 651–669.

Editors of Time-Life. 1992a. *Mass Murderers.* Alexandria, Va.: Time-Life Books.

———. 1992b. *Serial Killers.* Alexandria, Va.: Time-Life Books.

Ehrhardt, Anke A., and Susan W. Baker. 1978. "Fetal Androgens, Human Central Nervous System Differentiation, and Behavior Sex Differences." In *Sex Differences in Behavior.* Edited by Richard C. Friedman, Ralph M. Richart, and Raymond L. Vande Wiele. Huntington, N.Y.: Robert E. Krieger: 33–51.

Eibl-Eibesfeldt, Irenäus. 1989. *Human Ethology.* New York: Aldine de Gruyter.

Eldredge, Niles. 1982. *The Monkey Business: A Scientist Looks at Creationism.* New York: Washington Square.

Ember, Carol R. 1978. "Myths About Hunter-Gatherers." *Ethnology* 27: 239–448.

Esposito, John L. 1991. *Islam: The Straight Path.* Oxford: Oxford University Press.

Estep, Daniel Q., and Katherine E. M. Bruce. 1981. "The Concept of Rape in Animals: A Critique." *Animal Behaviour* 29: 1272–1273.

Evans, Robert R. 1969. *Readings in Collective Behavior.* Chicago: Rand McNally.

Ewald, Paul W. 1985. "Influence of Asymmetries in Resource Quality and Age on Aggression and Dominance in Black-chinned Hummingbirds." *Animal Behaviour* 33: 705–19.

Federal Bureau of Investigation. 1991. *Crime in the United States: Uniform Crime Reports.* Washington, D.C.: U.S. Government Printing Office.

"Femme Fatale." 1991. *New York Times* (February 2): 22.

Finley, Moses I. 1963. *The Ancient Greeks.* Harmondsworth, Middlesex, England: Penguin Books.

Fisher, Helen. 1992. *Anatomy of Love: The Natural History of Monogamy, Adultery, and Divorce.* New York: W. W. Norton.

Flowers, Ronald Barri. 1989. *Demographics and Criminality: The Characteristics of Crime in America.* New York: Greenwood Press.

Foley, Robert. 1987. *Another Unique Species: Patterns in Human Evolutionary Ecology.* Harlow, England: Longman.

Fortune, Reo F. 1939. "Arapesh Warfare." *American Anthropologist* 41: 22–41.

Fossey, Dian. 1983. *Gorillas in the Mist.* Boston: Houghton Mifflin.

Foster, Thomas W. 1975. "Make-Believe Families: A Response of Women and Girls to the Deprivations of Prison." *International Journal of Penology* 3: 71–78.

Fotheringham, Allan. 1991. "Can Wilt Really Count That High?" *MacLean's* (November 18): 84.

Frank, Laurence G. 1986. "Social Organization of the Spotted Hyaena (*Crocuta crocuta*). I. Demography." *Animal Behaviour* 35: 1500–1509.

Frank, Laurence G., Stephen E. Glickman, and Paul Licht. 1991. "Fatal Sibling Aggression, Precocial Development, and Androgens in Neonatal Spotted Hyenas." *Science* 252: 702–704.

Frank, Laurence G., Mary L. Weidele, and Stephen E. Glickman. 1995. "Masculinization Costs in Hyaenas." *Nature* 377: 584–585.

Freeman, Derek. 1964. "Human Aggression in Anthropological Perspective." In *The Natural History of Aggression*. Edited by J. D. Carthy and F. J. Ebling. New York: Academic Press: 109–119.

———. 1983. *Margaret Mead and Samoa: The Making and Unmaking of an Anthropological Myth*. Cambridge, Mass.: Harvard University Press.

Friedl, Ernestine. 1975. *Women and Men: An Anthropologist's View*. New York: Holt, Rinehart and Winston.

Fritz, Mark. 1994. "Confessions from a Massacre." *Boston Globe* (May 16): 1.

Furuichi, Takeshi. 1987. "Sexual Swelling, Receptivity, and Grouping of Wild Pygmy Chimpanzee Females at Wamba, Zaïre." *Primates* 28: 309–318.

———. 1989. "Social Interactions and the Life History of Female *Pan paniscus* in Wamba, Zaïre." *International Journal of Primatology* 10: 173–197.

———. 1992. "The Prolonged Estrus of Females and Factors Influencing Mating in a Wild Group of Bonobos (*Pan paniscus*) in Wamba, Zaïre." In *Topics in Primatology, vol. 2: Behavior, Ecology and Conservation*. Tokyo: University of Tokyo Press: 179–190.

Furuichi, Takeshi, and Hiroshi Ihobe. 1994. "Variation in Male Relationships in Bonobos and Chimpanzees." *Behaviour* 130: 211–228.

Galdikas, Biruté M. F. 1981. "Orangutan Reproduction in the Wild." In *Reproductive Biology of the Great Apes*. Edited by C. E. Graham. New York: Academic Press: 281–299.

———. 1995. *Reflections of Eden: My Years with the Orangutans of Borneo*. New York: Little, Brown.

Galili, U., and K. Swanson. 1991. "Gene Sequences Suggest Inactivation of alpha-1,3-galactosyltransferase in Catarrhines After the Divergence of Apes from Monkeys." *Proceedings of the National Academy of Sciences* 88: 7401–7404.

Garrett, Stephanie. 1987. *Gender*. London: Tavistock.

George, Kenneth M. 1991. "Headhunting, History and Exchange in Upland Sulawesi." *Journal of Asian Studies* 50: 536–564.

Gewertz, Deborah B. 1981. "A Historical Reconsideration of Female Dominance Among the Cambri of Papua New Guinea." *American Ethnologist* 8: 94–106.

Gewertz, Deborah B., and Frederick K. Errington. 1991. *Twisted Histories, Altered Contexts: Representing the Chambri in a World System.* Cambridge: Cambridge University Press.

Giallombardo, Rose. 1966. *Society of Women: A Study of a Women's Prison.* New York: Wiley.

Gibbs, Nancy R. 1995. "The Blood of Innocents." *Time* (May 1): 57–64.

Gilligan, Carol. 1982. *In a Different Voice: Psychological Theory and Women's Development.* Cambridge, Mass.: Harvard University Press.

Gilman, Charlotte Perkins. 1915; 1979. *Herland.* Introduced by Ann J. Lane. New York: Pantheon Books.

"Gingrich Calls Crimes 'Artifacts of Bad Policy.'" 1995. *Boston Globe* (May 20): 5.

Gish, Duane T. 1978. *Evolution: The Fossils Say No!* El Cajona, Calif.: Creation Life.

Gleick, Elizabeth. 1995. "Who Are They?" *Time* (May 1): 44–51.

Glickman, Stephen E., and others. 1992. "Sexual Differentiation of the Female Spotted Hyaena: One of Nature's Experiments." *Annals of the New York Academy of Sciences* 662: 135–159.

Goldfoot, D. A., and others. 1980. "Behavioral and Physiological Evidence of Sexual Climax in the Female Stump-tailed Macaque (*Macaca arctoides*)." *Science* 208: 1477–1479.

Golding, William. 1954. *Lord of the Flies.* New York: Putnam.

Goldman, Nancy Loring, editor. 1982. *Female Soldiers — Combatants or Noncombatants? Historical and Contemporary Perspectives.* Westport, Conn.: Greenwood Press.

Gonzales, I. L., and others. 1990. "Ribosomal RNA Sequences and Hominoid Phylogeny." *Molecular Biology and Evolution* 7: 203–219.

Goodall, Jane. 1986. *The Chimpanzees of Gombe: Patterns of Behavior.* Cambridge, Mass.: Harvard University Press.

———. 1990. *Through a Window.* Boston: Houghton Mifflin.

———. 1991. "Unusual Violence in the Overthrow of an Alpha Male Chimpanzee at Gombe." In *Topics in Primatology, Vol. 1: Human Origins.* Edited by Toshisada Nishida and others. Tokyo: University of Tokyo Press: 131–142.

Goodman, Morris. 1963. "Man's Place in the Phylogeny of the Primates as Reflected by Serum Proteins." In *Classification and Human Evolution.* Edited by S. L. Washburn. Chicago: Aldine: 204–235.

Gora, JoAnn Gennaro. 1982. *The New Female Criminal: Empirical Reality or Social Myth?* New York: Praeger.

Gould, Stephen Jay. 1995. "A Sea Horse for All Races." *Natural History* (November): 10–15, 72–75.

Gowaty, Patricia Adair. 1992. "Evolutionary Biology and Feminism." *Human Nature* 3 (3): 217–249.

Gray, Patrick, and Linda Wolfe. 1980. "Height and Sexual Dimorphism and Stature Among Human Societies." *American Journal of Physical Anthropology* 53: 441–456.

Greenfield, Les O. 1992. "Origin of the Human Canine." *Yearbook of Physical Anthropology* 35: 153–185.

Griesse, Anne Eliot, and Richard Stites. 1982. "Russia: Revolution and War." In *Female Soldiers — Combatants or Noncombatants? Historical and Contemporary Perspectives.* Edited by Nancy Loring Goldman. Westport, Conn.: Greenwood Press: 61–84.

Griffin, Susan. 1981. *Pornography and Silence: Culture's Revenge Against Nature.* New York: Harper and Row.

Grine, F. E., ed. 1988. *Robust Australopithecines.* Hawthorne, N.Y.: Aldine de Gruyter.

Gross, Daniel R. 1975. "Protein Capture and Cultural Development in the Amazon Basin." *American Anthropologist* 77 (3): 526–549.

Guérin, Daniel, ed. 1974; 1978. *The Writings of a Savage: Paul Gauguin.* Introduced by Wayne Anderson. New York: Viking.

Gulik, Robert H. van. 1974. *Sexual Life in Ancient China.* London: E. J. Brill.

Hammer, Joshua. 1994. "Escape from Hell." *Newsweek* (May 16): 34, 35.

Harrington, Fred. 1987. "The Man Who Cries Wolf." *Natural History* (February): 22–26.

Harris, Marvin. 1974. *Cows, Pigs, Wars and Witches: The Riddles of Culture.* New York: Random House.

———. 1979a. *Cultural Materialism: The Struggle for a Science of Culture.* New York: Random House.

———. 1979b. "The Yanomamö and the Causes of War in Band and Village Societies." In *Brazil: Anthropological Perspectives.* Edited by M. L. Margolis and W. E. Carter. New York: Columbia University Press: 121–133.

———. 1989. *Our Kind.* New York: Harper Collins.

Harvey, Paul H., Michael Kavanagh, and Timothy H. Clutton-Brock. 1978. "Canine Tooth Size in Female Primates." *Nature* 276: 817.

Hatley, Tom, and John Kappelman. 1980. "Bears, Pigs, and Plio-Pleistocene Hominids: A Case for the Exploitation of Belowground Food Resources." *Human Ecology* 8: 371–387.

Hausfater, Glenn, and Sarah Blaffer Hrdy, eds. 1984. *Infanticide: Comparative and Evolutionary Perspectives.* Hawthorne, N.Y.: Aldine de Gruyter.

Hayes, Cathy. 1951. *The Ape in Our House.* New York: Harper and Brothers.

Headland, Thomas N. 1992. *The Tasaday Controversy: Assessing the Evidence.* Washington, D.C.: American Anthropological Association.

Heise, Lori L., Jacqueline Pitanguy, and Adrienne Germain. 1994. *Violence Against Women: The Hidden Health Burden.* (World Bank Discussion Paper No. 255.) Washington, D.C.: The World Bank.

Herbert, T. Walter, Jr. 1980. *Marquesan Encounters: Melville and the Meaning of Civilization.* Cambridge, Mass.: Harvard University Press.

Heyes, C. 1995. "Self-recognition in Primates: Further Reflections Create a Hall of Mirrors." *Animal Behaviour* 50: 1533–1542.

Hitchens, Christopher. 1993. "Call of the Wilding." *Vanity Fair* (July): 30–35.

Hladik, C. Marcel, and Annette Hladik. 1990. "Food Resources of the Rain Forest." In *Food and Nutrition in the African Rain Forest.* Edited by C. M. Hladik, S. Bahuchet, and I. de Garine. Paris: UNESCO: 14–18.

Hofer, Heribert, and Marion L. East. 1993. "The Commuting System of Serengeti Spotted Hyaenas: How a Predator Copes with Migratory Prey. II. Intrusion Pressure and Commuters' Space Use." *Animal Behaviour* 46: 559–574.

Hohmann, Gotfried, and Barbara Fruth. 1993. "Field Observations on Meat Sharing Among Bonobos *(Pan paniscus).*" *Folia Primatologica* 60: 225–229.

Hölldobler, Bert. 1976. "Tournaments and Slavery in a Desert Ant." *Science* 192: 912–914.

———. 1981. "Foraging and Spatiotemporal Territories in the Honey Ant *Myrmecocystus mimicus* Wheeler (Hymenoptera: Formicidae)." *Behavioral Ecology and Sociobiology* 9: 301–314.

Hölldobler, Bert, and Charles J. Lumsden. 1980. "Territorial Strategies in Ants." *Science* 210: 732–739.

Holloway, Marguerite. 1994. "Trends in Women's Health: A Global View." *Scientific American* (August): 77–83.

Holmberg, Allan R. 1969. *Nomads of the Long Bow: The Siriono of Eastern Bolivia.* Garden City, N.Y.: Natural History Press.

Hooff, Jan A. R. A. M. van, and Carel P. van Schaik. 1994. "Male Bonds:

Affiliative Relationships Among Nonhuman Primate Males." *Behaviour* 130: 309–337.

Hooks, Bonnie L., and Penny A. Green. 1993. "Cultivating Male Allies: A Focus on Primate Females, Including Homo sapiens." *Human Nature* 4 (1).

Horai, Satoshi, and others. 1992. "Man's Place in Hominoidea Revealed by Mitochondrial DNA Geneaology." *Journal of Molecular Evolution* 35: 32–43.

———. 1995. "Recent African Origin of Modern Humans Revealed by Complete Sequences of Hominoid Mitochondrial DNAs." *Proceedings of the National Academy of Sciences* 92: 532–536.

Hornblower, Margot. 1993. "The Skin Trade." *Time* (June 21): 45–51.

Howard, Michael. 1983a. "The Causes of War." In *The Causes of War and Other Essays*. London: Unwin Paper: 7–22.

———. 1983b. *Clausewitz*. Oxford: Oxford University Press.

Hrdy, Sarah Blaffer. 1981. *The Woman that Never Evolved*. Cambridge, Mass.: Harvard University Press.

Hrdy, Sarah Blaffer, Charles Janson, and Carel van Schaik. 1995. "Infanticide: Let's Not Throw Out the Baby with the Bathwater." *Evolutionary Anthropology* 3: 151–154.

Hunt, Kevin D. 1989. "Positional Behavior in *Pan troglodytes* at the Mahale Mountains and the Gombe Stream National Parks, Tanzania." Ph.D. dissertation: University of Michigan, Ann Arbor.

———. 1994. "The Evolution of Human Bipedality: Ecology and Functional Morphology." *Journal of Human Evolution* 26: 183–202.

Huntingford, Felicity A. 1976. "The Relationship between Intra- and Inter-Specific Aggression." *Animal Behaviour* 24: 485–497.

Huxley, Thomas H. 1863; 1894. *Man's Place in Nature*. New York: D. Appleton.

Idani, Gen'ichi. 1991a. "Cases of Inter-unit Group Encounters in Pygmy Chimpanzees at Wamba, Zaïre." In *Primatology Today: Proceedings of the XIIIth Congress of the International Primatological Society*. Edited by Akiyoshi Ehara and others. Amsterdam: Elsevier: 235–238.

———. 1991b. "Social Relationships Between Immigrant and Resident Bonobo *(Pan paniscus)* Females at Wamba." *Folia Primatologica* 57: 83–95.

Ihobe, Hiroshi. 1990. "Interspecific Interactions Between Wild Pygmy Chimpanzees *(Pan paniscus)* and Red Colobus *(Colobus badius)*." *Primates* 31: 109–112.

————. 1992. "Observations on the Meat-eating Behavior of Wild Bonobos *(Pan paniscus)* at Wamba, Republic of Zaïre." *Primates* 33: 247–250.

Ingrassia, Michele, and Melinda Beck. 1994. "Patterns of Abuse." *Newsweek* (July 4): 26–33.

Jamison, Paul L. 1978. "Anthropometric Variation." In *Eskimos of Northwestern Alaska*. Edited by P. L. Jamison, S. L. Zegura, and F. A. Milan. Stroudsberg, Penn.: Dowden, Hutchinson, and Ross: 40–78.

Janson, Charles H., and Michele L. Goldsmith. 1995. "Predicting Group Size in Primates: Foraging Costs and Predation Risks." *Behavioral Ecology* 6 (3): 326–336.

Jarman, Peter J. 1989. "Sexual Dimorphism in Macropodoidea." In *Kangaroos, Wallabies and Rat-kangaroos*. Edited by G. Grigg, Peter Jarman, and Ian Hume. New South Wales, Australia: Surrey Beatty and Sons: 433–447.

Jarvis, Jennifer U. M. 1994. "Mammalian Eusociality: A Family Affair." *Trends in Ecology and Evolution* 9: 47–51.

Jones, Steve, Robert Martin, and David Pilbeam. 1992. *The Cambridge Encyclopaedia of Human Evolution*. Cambridge: Cambridge University Press.

Jones, Steven. 1994. *The Language of Genes: Solving the Mysteries of Our Genetic Past, Present and Future*. New York: Anchor Books.

Joubert, Dereck. 1994. "Lions of Darkness." *National Geographic* 186: 35–53.

Joubert, Dereck, and Beverly Joubert. 1992. *Eternal Enemies*. A film produced by Wildlife Films Botswana.

Kano, Takayoshi. 1979. "A Pilot Study on the Ecology of Pygmy Chimpanzees *Pan paniscus*." In *The Great Apes*. Edited by D. A. Hamburg and E. R. McCrown. Menlo Park, Calif.: Benjamin/Cummings: 123–136.

————. 1982. "The Social Group of Pygmy Chimpanzees *(Pan paniscus)* of Wamba." *Primates* 23: 171–188.

————. 1990. "The Bonobo's Peaceable Kingdom." *Natural History* (November): 62–71.

————. 1992. *The Last Ape: Pygmy Chimpanzee Behavior and Ecology*. Translated by Evelyn Ono Vineberg. Stanford, Calif.: Stanford University Press.

Karl, Pierre. 1991. *Animal and Human Aggression*. New York: Oxford University Press.

Keegan, John. 1993. *A History of Warfare*. New York: Alfred A. Knopf.

Keeley, Lawrence H. 1996. *War Before Civilization: The Myth of the Peaceful Savage*. Oxford: Oxford University Press.

Keirans, J. E. 1984. *George Henry Falkiner Nuttall and the Nuttall Tick Catalogue*. Washington, D.C.: U.S. Government Printing Office.

Kingsley, Susan. 1988. "Physiological Development of Male Orangutans and Gorillas." In *Orang-utan Biology*. Edited by Jeffrey H. Schwartz. Oxford: Oxford University Press: 123–131.

Kinzey, W. G. 1984. "The Dentition of the Pygmy Chimpanzee, *Pan paniscus*." In *The Pygmy Chimpanzee: Evolutionary Biology and Behavior*. Edited by R. L. Susman. New York: Plenum: 65–88.

Klare, Hugh J. 1979. *Anatomy of Prison*. Westport, Conn.: Greenwood Press.

Klein, Richard G. 1989. *The Human Career: Human Biological and Cultural Origins*. Chicago: University of Chicago Press.

Knauft, Bruce M. 1987. "Reconsidering Violence in Simple Human Societies: Homicide Among the Gebusi of New Guinea." *Current Anthropology* 28: 457–500

———. 1991. "Violence and Sociality in Human Evolution." *Current Anthropology* 32 (4) (August–October): 391–409.

Kortlandt, Adriaan. 1972. *New Perspectives on Ape and Human Evolution*. Amsterdam: Stichting voor Psychobiologie.

———. 1993. "The Discovery of the Pygmy Chimpanzee: In 1913!" *Bulletin of the American Society of Primatologists* 17 (1).

Kruttschnitt, Candace. 1981. "Prison Codes, Inmate Solidarity, and Women: A Reexamination." In *Comparing Female and Male Offenders*. Edited by Marguerite Q. Warren. Beverly Hills, Calif.: Sage Publications: 123–144.

Kruuk, Hans. 1972. *The Spotted Hyena: A Study of Predation and Social Behavior*. Chicago: University of Chicago Press.

Kuntz, Tom. 1995. "Rhett and Scarlett: Rough Sex or Rape? Feminists Give a Damn." *New York Times* (February 19): E7.

Kuper, Adam. 1994. *The Chosen Primate: Human Nature and Cultural Diversity*. Cambridge, Mass.: Harvard University Press.

Kuroda, Suehisa. 1979. "Grouping of the Pygmy Chimpanzees." *Primates* 20: 161–183.

Lang, Gretchen. 1994. "Rwandan Mission Provides No Refuge." *Boston Globe* (June 1): 2.

Las Casas, Bartolomé de. 1542; 1992. *The Devastation of the Indies: A*

Brief Account. Translated by H. Briffault. Baltimore, Md.: Johns Hopkins University Press.

Law, Robin. 1993. "The 'Amazons' of Dahomey." *Paideuma* 39: 246–260.

Leacock, Eleanor. 1993. "Women in Samoan History: A Further Critique of Derek Freeman." In *Sex and Gender Hierarchies.* Edited by Barbara Diane Miller. Cambridge: Cambridge University Press: 351–365.

Lee, Richard B. 1979. *The !Kung San: Men, Women, and Work in a Foraging Society.* Cambridge: Cambridge University Press.

———. 1982. "Politics, Sexual and Non-sexual, in an Egalitarian Society." In *Politics and History in Band Societies.* Edited by Eleanor Leacock and Richard Lee. Cambridge: Cambridge University Press: 37–59.

Leland, John, and others. 1995. "Why the Children?" *Newsweek* (May 1): 48–53.

Leonard, Eileen B. 1982. *Women, Crime, and Society: A Critique of Theoretical Criminology.* New York: Longman.

Lerner, Gerda. 1986. *The Creation of Patriarchy.* New York: Oxford University Press.

Levinson, David. 1989. *Family Violence in Cross-Cultural Perspective.* Newbury Park, Calif.: Sage Publications.

Levy, Robert I. 1983. "The Attack on Mead." *Science* (May 20): 829–832.

Lewin, Roger. 1993. *Human Evolution: An Illustrated Introduction.* Oxford: Blackwell Scientific.

Lorch, Donatella. 1994. "Bodies from Rwanda Cast a Pall on Lakeside Villages in Uganda." *New York Times* (May 28): 1, 5.

Lorenz, Konrad. 1963; 1966. *On Aggression.* Translated by Marjorie Latzke. London: Methuen.

MacDonald, Eileen. 1991. "Female Terrorists." *Marie Claire* (July): 47 ff.

MacKinnon, John. 1971. "The Orang-utan in Sabah Today." *Oryx* 11 (May): 141–91.

———. 1974. "The Behaviour and Ecology of Wild Orang-utans *(Pongo pygmaeus).*" *Animal Behaviour* 22: 3–74.

Malenky, Richard K., and others. 1994. "The Significance of Terrestrial Herbaceous Foods for Bonobos, Chimpanzees and Gorillas." In *Chimpanzee Cultures.* Edited by Richard W. Wrangham and others. Cambridge, Mass.: Harvard University Press: 59–75.

Malenky, Richard K., and Richard W. Wrangham. 1994. "A Quantitative Comparison of Terrestrial Herbaceous Food Consumption by *Pan paniscus* in the Lomako Forest, Zaïre, and *Pan troglodytes* in the Kibale Forest, Uganda." *American Journal of Primatology* 32: 1–12.

Malina, R. M., and C. Bouchard. 1991. *Growth, Maturation and Physical Activity.* Champaign-Urbana: Human Kinetics Books.

Mallia, Joseph. 1995. "Militia Is Fighting Mad." *Boston Herald* (April 30): 3.

Manson, Joseph H., and Richard W. Wrangham. 1991. "Intergroup Aggression in Chimpanzees and Humans." *Current Anthropology* 32 (4): 369–390.

Marks, Jonathan. 1993. "Hominoid Heterochromatin: Terminal C-Bands as a Complex Genetic Trait Linking Chimpanzee and Gorilla." *American Journal of Physical Anthropology* 90: 237–249.

Marks, Jonathan, Carl W. Schmid, and Vincent M. Sarich. 1988. "DNA Hybridization as a Guide to Phylogeny: Relationships of the Hominoidea." *Journal of Human Evolution* 17: 769–786.

Marx, Leo. 1964. *The Machine in the Garden: Technology and the Pastoral Ideal in America.* Oxford: Oxford University Press.

McHenry, Henry M. 1994. "Behavioral Ecological Implications of Early Hominid Body Size." *Journal of Human Evolution* 27: 77–87.

Mead, Margaret. 1928a. *Coming of Age in Samoa: A Psychological Study of Primitive Youth for Western Civilization.* New York: William Morrow.

———. 1928b. "The Rôle of the Individual in Samoan Culture." *Journal of the Royal Anthropological Institute* 58: 487.

———. 1935; 1963. *Sex and Temperament in Three Primitive Societies.* New York: William Morrow.

———. 1940. "Social Change and Cultural Surrogates." *Journal of Educational Sociology* 14 (2) (October): 92–109.

———. 1949. *Male and Female: A Study of the Sexes in a Changing World.* New York: William Morrow.

———. 1972. *Blackberry Winter: My Earlier Years.* New York: William Morrow.

———. 1977. *Letters from the Field, 1925–1975.* New York: Harper and Row.

Mech, L. David. 1977. "Productivity, Mortality and Population Density of Wolves in Northeastern Minnesota." *Journal of Mammology* 58: 559–574.

Meggitt, Mervyn. 1977. *Blood Is Their Argument: Warfare Among the Mae Enga Tribesmen of the New Guinea Highlands.* Palo Alto, Calif.: Mayfield Publishing.

Melville, Herman. 1846; 1972. *Typee: A Peep at Polynesian Life.* Edited by George Woodcock. London: Penguin.

Mernissi, Fatima. 1994. *Dreams of Trespass: Tales of a Harem Girlhood.* Reading, Mass.: Addison-Wesley.

Mesquita, Bruce Bueno de. 1981. *The War Trap.* New Haven: Yale University Press.

Minderhout, D. T. 1986. "Introductory Texts and Social Sciences Stereotypes." *Anthropology Newsletter* 27 (3): 14–15.

Mitani, John. 1985. "Mating Behaviour of Male Orangutans in the Kutai Game Reserve, Indonesia." *Animal Behaviour* 33: 392–402.

Mitchell, Margaret. 1936. *Gone with the Wind.* New York: Macmillan.

Moore, Jim. 1992. "'Savanna' Chimpanzees." In *Topics in Primatology, vol. 1: Human Origins.* Edited by Toshisada Nishida and others. Tokyo: University of Tokyo Press: 99–118.

Moorehead, Alan. 1960. *The White Nile.* New York: Harper and Row.

———. 1969. *Darwin and the Beagle.* New York: Harper and Row.

Morell, Virginia. 1995. "Chimpanzee Outbreak Heats Up Search for Ebola Origin." *Science* 268: 974, 975.

Morgan, John. 1852; 1979. *The Life and Adventures of William Buckley: Thirty-Two Years a Wanderer Amongst the Aborigines.* Canberra: Australia National University Press.

Morin, Phillip A., and others. 1994. "Kin Selection, Social Structure, Gene Flow, and the Evolution of Chimpanzee." *Science* 265: 1193–1201.

Morganthau, Tom, and others. 1995. "The View from the Far Right." *Newsweek* (May 1): 36–39.

Muroyama, Y., and Y. Sugiyama. 1994. "Grooming Relationships in Two Species of Chimpanzees." In *Chimpanzee Cultures.* Edited by Richard W. Wrangham and others. Cambridge, Mass.: Harvard University Press: 169–180.

Murray, Martyn G., and Russell Gerrard. 1984. "Conflicts in the Neighbourhood: Models Where Close Relatives Are in Direct Competition." *Journal of Theoretical Biology* 111: 237–46.

Nance, John. 1975. *The Gentle Tasaday: A Stone Age People in the Philippine Rain Forest.* New York: Harcourt Brace Jovanovich.

Nash, Robert Jay. 1973. *Bloodletters and Badmen: A Narrative Encyclo-*

pedia of American Criminals from the Pilgrims to the Present. New York: M. Evans.

Nishida, Toshisada, ed. 1990. *The Chimpanzees of the Mahale Mountains: Sexual and Life History Strategies.* Tokyo: University of Tokyo Press.

Nishida, Toshisada, Mariko Haraiwa-Hasegawa, and Yuko Takahata. 1985. "Group Extinction and Female Transfer in Wild Chimpanzees in the Mahale National Park, Tanzania." *Zeitschrift für Tierpsychologie* 67: 284–301.

Nishida, Toshisada, and Kenji Kawanaka. 1985. "Within-Group Cannibalism by Adult Male Chimpanzees." *Primates* 26: 274–284.

Nishida, Toshisada, and others. 1991. "Meat-Sharing as a Coalition Strategy by an Alpha Male Chimpanzee?" In *Topics in Primatology, vol. 1: Human Origins.* Edited by Toshisada Nishida and others. Tokyo: University of Tokyo Press: 159–176.

Nuttall, George H. F. 1904. *Blood Immunity and Blood Relationship: A Demonstration of Certain Blood-relationships Amongst Animals by Means of the Precipitin Test for Blood.* Cambridge: Cambridge University Press.

O'Malley, Suzanne. 1993. "Girlz 'n the Hood." *Harper's Bazaar* (October): 238 ff.

Ortner, Sherry B. 1974. "Is Female to Male as Nature Is to Culture?" In *Woman, Culture and Society.* Edited by Michelle Z. Rosaldo and Louise Lamphere. Stanford, Calif.: Stanford University Press: 67–87.

Otterbein, Keith F. 1970. *The Evolution of War.* New Haven: HRAF Press.

Packer, Craig, and Anne Pusey. 1983. "Adaptations of Female Lions to Infanticide by Incoming Males." *American Naturalist* 121: 716–728.

Pagnozzi, Amy. 1994. "Killer Girls." *Elle* (May): 122–126.

Palmer, Craig T. 1989a. "Is Rape a Cultural Universal?: A Re-examination of the Ethnograhic Data." *Ethnology* 28: 1–16.

———. 1989b. "Rape in Nonhuman Animal Species: Definitions, Evidence, and Implications." *Journal of Sex Research* 26 (3): 355–374.

———. 1991. "Human Rape: Adaptation or By-product?" *Journal of Sex Research* 28 (3): 365–386.

Parish, A. R. 1993. "Sex and Food Control in the 'Uncommon Chimpanzee': How Bonobo Females Overcome a Phylogenetic Legacy of Male Dominance." *Ethology and Sociobiology* 15 (3): 157–179.

Parker, Geoffrey A. 1974. "Assessment Strategy and the Evolution of Fighting Behavior." *Journal of Theoretical Biology* 47: 223–243.

Parker, Gary. 1980. *Creation: The Facts of Life.* El Cajon, Calif.: Creation Life.

Parmigiani, Stefano, and Frederick S. vom Saal, eds. 1994. *Infanticide and Parental Care.* London: Ettore Majorana Life Sciences Series, Harwood Academic.

Percival, L., and K. Quinkert. 1987. "Anthropometric Factors." In *Sex Differences in Human Performance.* Edited by Mary Baker. New York: John Wiley and Sons: 121–139.

Perkins, Roberta, and Garry Bennett. 1985. *Being a Prostitute: Prostitute Women and Prostitute Men.* London: George Allen and Unwin.

Peters, Charles, and Eileen O'Brien. 1994. "Potential Hominid Plant Foods from Woody Species in Semi-arid Versus Sub-humid Subtropical Africa." In *The Digestive System in Mammals: Food and Function.* Edited by D. J. Chivers and P. Langer. Cambridge: Cambridge University Press: 166–192.

Peterson, Dale, and Jane Goodall. 1993. *Visions of Caliban: On Chimpanzees and People.* Boston: Houghton Mifflin.

Pierre-Pierre, Garry. 1995. "Two Women Charged with Series of Armed Robberies in Brooklyn." *New York Times* (January 26): B1, B3.

Pilbeam, David. 1995. "Genetic and Morphological Records of the Hominoidea and Hominid Origins: A Synthesis." *Molecular Phylogenetics and Evolution* (in press).

Plavcan, J. Michael, and others. 1995. *Journal of Human Evolution* 28: 245–276.

Plavcan, J. Michael, and Carel P. van Schaik. 1992. "Intrasexual Competition and Canine Dimorphism in Anthropoid Primates." *American Journal of Physical Anthropology* 87: 461–477.

———. 1994. "Canine Dimorphism." *Evolutionary Anthropology* 2: 208–214.

Pollitt, Katha. 1992. "Are Women Morally Superior to Men?" *The Nation* (December 28): 799–807.

Pollock-Byrne, Jocelyn M. 1990. *Women, Prison, and Crime.* Pacific Grove, Calif.: Brooks/Cole.

Popp, Joseph, and Irven De Vore. 1979. "Aggressive Competition and Social Dominance Theory." In *The Great Apes.* Edited by D. A. Hamburg and E. R. McCrown. Menlo Park, Calif.: Benjamin/Cummings: 317–338.

Porter, Roy. 1986. "Rape — Does It Have a Historical Meaning?" In *Rape*. Edited by Sylvana Tomaselli and Roy Porter. Oxford: Basil Blackwell: 216–236.

Power, Margaret. 1991. *The Egalitarians, Human and Chimpanzee: An Anthropological View of Social Organization*. Cambridge: Cambridge University Press.

Putnam, Robert. 1993. *Making Democracy Work: Civic Traditions in Modern Italy*. Princeton: Princeton University Press.

Rabbie, Jacob M. 1992. "The Effects of Intragroup Cooperation and Intergroup Competition on In-group Cohesion and Out-group Hostility." In *Coalitions and Alliances in Humans and Other Animals*. Edited by Alexander H. Harcourt and Frans B. M. de Waal. Oxford: Oxford University Press: 175–205.

Raghavan, Sudarsan, Hassan Shahriar, and Fazal Qureshi. 1994. "Warriors of God." *Newsweek* (August 15): 36, 37.

Raper, A. F. 1933. *The Tragedy of Lynching*. Durham: University of North Carolina Press.

"Ready for War: Inside the World of the Paranoid." 1995. *New York Times* (April 30): 1, E5.

Reichert, Susan E. 1978. "Games Spiders Play: Behavioral Variability in Territorial Disputes." *Behavioral Ecology and Sociobiology* 3: 135–162.

Reinisch, June Machover, Mary Ziemba-Davis, and Stephanie A. Sanders. 1991. "Hormonal Contributions to Sexual Dimorphic Behavioral Development in Humans." *Psychoneuroendocrinology* 16 (1–3): 213–278.

Reynolds, Vernon. 1967. *The Apes*. New York: Harper Colophon.

Richardson, Lewis F. 1960. *The Statistics of Deadly Quarrels*. Pittsburgh: Boxwood.

"Ricki Lake on Line Two, Mr. Z." 1995. *Time* (July 3): 61.

Rijksen, Herman D. 1978. *A Field Study of Sumatran Orangutans*. Wageningen, Netherlands: H. Veenman and B. V. Zonen.

———. 1974. "Social Structure in a Wild Orang-utan Population in Sumatra." In *Contemporary Primatology*. Edited by S. Kondo and others. Basel: S. Karger: 373–379.

Robarchek, Clayton A., and Carole J. Robarchek. 1992. "Cultures of War and Peace: A Comparative Study of Waorani and Semai." In *Aggression and Peacefulness in Humans and Other Primates*. Edited by James Silverberg and J. Patrick Gray. New York: Oxford University Press: 189–213.

Rodseth, Lars, and others. 1991. "The Human Community as a Primate Society." *Current Anthropology* 32 (3) (June): 221–254.

Rodman, Peter, and John Mitani. 1987. "Orangutans: Sexual Dimorphism in a Solitary Species." In *Primate Societies*. Edited by Barbara B. Smuts and others. Chicago: Chicago University Press: 146–154.

Rubenstein, Charles A. 1974. "Polygamy." In *Universal Jewish Encyclopedia*, vol. 8. New York: UJE: 584, 585.

Ruddick, Sara. 1990. *Maternal Thinking: Toward a Politics of Peace.* New York: Ballantine.

Ruvolo, Maryellen, and others. 1991. "Resolution of the African Hominoid Trichotomy by Use of a Mitochondrial Gene Sequence." *Proceedings of the National Academy of Sciences* 88: 1570–1574.

———. 1994. "Gene Trees and Hominoid Phylogeny." *Proceedings of the National Academy of Sciences* 91: 8900–8904.

Sabater Pi, Jorge, and others. 1993. "Behavior of Bonobos *(Pan paniscus)* Following Their Capture of Monkeys in Zaïre." *International Journal of Primatology* 14: 797–804.

Sanday, Peggy Reeves. 1981. "A Socio-Cultural Context of Rape: A Cross-Cultural Study." *Journal of Social Issues* 37: 5–27.

———. 1986. "Rape and the Silencing of the Feminine." In *Rape*. Edited by Sylvana Tomaselli and Roy Porter. New York: Basil Blackwell: 84–101.

Sapolsky, Robert M. 1994. *Why Zebras Don't Get Ulcers: A Guide to Stress, Stress-related Diseases, and Coping.* New York: W. H. Freeman.

Savage, Robert J. G. 1988. "Extinction and the Fossil Mammal Record." In *Extinction and Survival in the Fossil Record.* Edited by G. P. Larwood. Oxford: Clarendon: 319–334.

Savage-Rumbaugh, Sue, and Roger Lewin. 1994. *Kanzi: The Ape at the Brink of the Human Mind.* New York: Wiley.

Schaik, Carel van, and Robin Dunbar. 1990. "The Evolution of Monogamy in Large Primates: A New Hypothesis and Some Crucial Tests." *Behaviour* 115: 30–61.

Schaik, Carel van, and Peter Kappeler. 1993. "Life History, Activity Period and Lemur Social Systems." In *Lemur Social Systems and Their Ecological Basis*. Edited by Peter Kappeler and Jorge Ganzhorn. New York: Plenum Press: 241–260.

Schaller, George. 1972. *The Serengeti Lion: A Study of Predator-Prey Relations.* Chicago: University of Chicago Press.

Schwartz, Jeffrey H. 1984. "The Evolutionary Relationships of Man and the Orangutans." *Nature* 308: 501–505.

————. 1987. *The Red Ape: Orang-utans and Human Origins.* Boston: Houghton Mifflin.

Schwarz, Ernst. 1929. "Das Vorkommen des Schimpansen auf den Linken Congo-Ufer." *Revue de Zoologie et de Botanique Africaines* 16 (4): 425–426.

Shabad, Steven. 1995. "Beyond the Fringe." *Newsweek* (May 1): 38.

Shakur, Sanyika. 1993. *Monster: The Autobiography of an L.A. Gang Member.* New York: Atlantic Monthly Press.

Shea, Brian T. 1983. "Phyletic Size Change and Brain/Body Allometry: A Consideration Based on the African Pongids and Other Primates." *International Journal of Primatology* 4: 33–62.

————. 1984. "Between the Gorilla and the Chimpanzee: A History of Debate Concerning the Existence of the *Kooloo-Kamba* or Gorilla-like Chimpanzee." *Journal of Ethnobiology* 4: 1–13.

————. 1985. "Ontogenetic Allometry and Scaling: A Discussion Based on Growth and Form of the Skull in African Apes." In *Size and Scaling in Primate Biology.* Edited by William L. Jungers. New York: Plenum: 175–205.

Sherif, Muzafer, and others. 1961. *Intergroup Conflict and Cooperation: The Robber's Cave Experiment.* Norman, Okla.: University of Oklahoma Book Exchange.

Short, R. V. 1980. "The Great Apes of Africa." *Journal of Reproductive Fertility,* Supplement 28: 3–11.

Shostak, Marjorie. 1981. *Nisa: The Life and Words of a !Kung Woman.* Cambridge, Mass.: Harvard University Press.

Shoumatoff, Alex. 1994. "Flight from Death." *The New Yorker* (June 20): 44–55.

Sibley, Charles G., and Jon E. Ahlquist. 1983. "The Phylogeny and Classification of Birds Cased on the Data of DNA-DNA Hybridization." In *Current Ornithology,* vol 1. Edited by R. F. Johnston. New York: Plenum: 245–292.

————. 1984. "The Phylogeny of the Hominid Primates, as Indicated by DNA-DNA Hybridization." *Journal of Molecular Evolution* 20: 2–15.

Sibley, Charles G., J. A. Comstock, and Jon E. Ahlquist. 1990. "DNA Hybridization Evidence of Hominoid Phylogeny: A Reanalysis of the Data." *Journal of Molecular Evolution* 30: 202–236.

Sikes, Gini. 1994. "Girls in the 'Hood." *Scholastic Update* (February 11): 20 ff.

Silk, Joan B. 1993. "Primatological Perspectives on Gender Hierarchies." In *Sex and Gender Hierarchies.* Edited by Barbara Diane Miller. Cambridge: Cambridge University Press: 212–235.

Sillen, Andrew. 1992. "Strontium-calcium Ratios (Sr/Ca) of *Australopithecus robustus* and Associated Fauna from Swartkrans." *Journal of Human Evolution* 23: 495-516.

Simmel, Georg. 1969. "The Individual and the Mass." In *Readings in Collective Behavior*, 1st ed. Edited by Robert R. Evans. Chicago: Rand McNally: 39–45.

Simon, Rita James. 1975. *Women and Crime.* Lexington, Mass.: Lexington Books.

Small, Melvin, and J. David Singer. 1983. *Resort to Arms: International and Civil Wars, 1816–1980.* Beverly Hills, Calif.: Sage Publications.

Small, Meredith. 1993. *Female Choices: Sexual Behavior of Female Primates.* Ithaca, N.Y.: Cornell University Press.

Smith, Robert L. 1984. "Human Sperm Competition." In *Sperm Competition and the Evolution of Animal Mating Systems.* Edited by Robert L. Smith. New York: Academic Press: 601–659.

Smuts, Barbara B. 1985. *Sex and Friendship in Baboons.* New York: Aldine de Gruyter.

———. 1992. "Male Aggression Against Women: An Evolutionary Perspective." *Human Nature* 3: 1–44.

———. 1995. "The Evolutionary Origins of Patriarchy." *Human Nature*, 1: 1–32.

Smuts, Barbara B., and Robert W. Smuts. 1993. "Male Aggression and Sexual Coercion of Females in Nonhuman Primates and Other Animals: Evidence and Theoretical Implications." *Advances in the Study of Behavior* 22: 1–63.

Sorokin, Pitirim A. 1962. *Social and Cultural Dynamics*, vol. 3. New York: Bedminster Press.

Spaeth, Anthony. 1995. "Engineer of Doom." *Time* (June 12): 57.

Stanford, Craig B., and others. 1994. "Patterns of Predation by Chimpanzees on Red Colobus Monkeys in Gombe National Park, 1982–1991." *American Journal of Physical Anthropology* 94: 213-228.

Stanley, Steven M. 1992. "An Ecological Theory for the Origin of Homo." *Paleobiology* 18: 237–257.

Starin, E. Dawn. 1994. "Philopatry and Affiliation Among Red Colobus." *Behaviour* 130: 253–269.

Stauffer, John, and Richard Frost. 1976. "Male and Female Interest in Sexually-Oriented Magazines." *Journal of Communication* 26 (Winter): 25–30.

Stephan, C. W., and W. G. Stephan. 1990. *Two Social Psychologies*, 2nd ed. Belmont, Calif.: Wadsworth.

Stocking, George W., Jr. 1989. "The Ethnographic Sensibility of the 1920s and the Dualism of the Anthropological Tradition." In *Romantic Motives: Essays on Anthropological Sensibility. History of Anthropology*, vol. 6. Edited by George W. Stocking, Jr. Madison, Wis.: University of Wisconsin Press: 208–276.

Strier, Karen B. 1992. *Faces in the Forest: The Endangered Muriqui Monkeys of Brazil.* Oxford: Oxford University Press.

———. 1994. "Brotherhoods Among Atelins: Kinship, Affiliation and Competition." *Behaviour* 130: 151–167.

Sugiyama, Yukimaru. 1988. "Grooming Interactions Among Adult Chimpanzees at Bossou, Guinea, with Special Reference to Social Structure." *International Journal of Primatology* 9: 393–407.

Sumner, W. G. 1906. *Folkways*. Boston: Ginn.

Susman, Randall L. 1988. "Hand of *Paranthropus robustus* from Member 1, Swartkrans: Fossil Evidence for Tool Behavior." *Science* 240: 781–784.

Sussman, Robert W., James M. Cheverud, and Thad Q. Bartlett. 1994. "Infant Killing as an Evolutionary Strategy: Reality or Myth?" *Evolutionary Anthropology*, vol. 3, no. 5: 149–151.

Swiss, Shana, and Joan E. Giller. 1993. "Rape as a Crime of War: A Medical Perspective." *Journal of the American Medical Association* 270: 612–615.

Symons, Donald. 1979. *The Evolution of Human Sexuality*. Oxford: Oxford University Press.

Takahata, Naoyuki. 1995. "A Genetic Perspective on the Origin of Man and His History." *Annual Review of Ecology and Systematics* (in press).

Tannen, Deborah. 1986. *That's Not What I Meant: How Conversational Style Makes or Breaks Relationships.* New York: William Morrow.

———. 1990. *You Just Don't Understand: Men and Women in Conversation.* New York: William Morrow.

Tanner, James M. 1978. *Fetus into Man: Physical Growth from Conception to Maturity.* Cambridge, Mass.: Harvard University Press.

Taylor, Carl S. 1990. *Dangerous Society.* East Lansing, Mich.: Michigan State University Press.

Tempkin, Jennifer. 1986. "Women, Rape and Law Reform." In *Rape.* Edited by Sylvana Tomaselli and Roy Porter. New York: Basil Blackwell: 16–40.

Terborgh, John. 1992. *Diversity and the Tropical Rain Forest.* New York: W. H. Freeman.

Thomas, Evan, and others. 1995. "Cleverness — And Luck." *Newsweek* (May 1): 35.

Thompson-Handler, Nancy, Richard Malenky, and Gay Reinharz. 1995. *Action Plan for* Pan paniscus: *Report on Free Ranging Populations and Proposals for Their Preservation.* Milwaukee, Wis.: The Zoological Society of Milwaukee County.

Thornhill, Randy. 1979. "Male and Female Sexual Selection and the Evolution of Mating Strategies in Insects." In *Sexual Selection and Reproductive Competition in Insects.* Edited by Murray S. Blum and Nancy A. Blum. New York: Academic Press: 81–121.

———. 1980. "Rape in *Panorpa* Scorpionflies and a General Rape Hypothesis." *Animal Behaviour* 28: 52–59.

Thornhill, Randy, and Nancy W. Thornhill. 1992. "The Evolutionary Psychology of Men's Coercive Sexuality." *Behavior and Brain Sciences* 15: 363–421.

Thornhill, Randy, Nancy W. Thornhill, and Gerard A. Dizinno. 1986. "The Biology of Rape." In *Rape.* Edited by Sylvana Tomaselli and Roy Porter. New York: Basil Blackwell: 102–121.

Tiger, Lionel. 1969. *Men in Groups.* New York: Random House.

———. 1987. *The Manufacture of Evil: Ethics, Evolution and the Industrial System.* New York: Harper and Row.

Tonkinson, Robert. 1988. "Ideology and Domination in Aboriginal Australia: A Western Desert Test Case." In *Hunters and Gatherers 2: Property, Power and Ideology.* Edited by Tim Gold, Davis Riches, and James Woodburn. New York: St. Martin's Press.

Tratz, E., and H. Heck. 1954. "Der Afrikanische Anthropoide 'Bonobo,' eine neue Menschenaffengattung." *Säugetierkundliche Mitteilungen* 2: 97–101.

Trinkaus, Erik. 1978. "Hard Times Among the Neanderthals." *Natural History* 87 (10): 58–63.

Turnbull, Colin M. 1965. *Wayward Servants: The Two Worlds of African Pygmies.* Garden City, N.Y.: Natural History Press.

———. 1982. "The Ritualization of Potential Conflict Between the Sexes Among the Mbuti." In *Politics and History in Band Societies.*

Edited by Eleanor Leacock and Richard Lee. Cambridge: Cambridge University Press: 133–155.

Turney-High, Harry H. 1949; 1991. *Primitive War: Its Practice and Concepts.* Columbia, S.C.: University of South Carolina Press.

Tuten, Jeff M. 1982. "Germany and the World Wars." In *Female Soldiers — Combatants or Noncombatants? Historical and Contemporary Perspectives.* Edited by Nancy Loring Goldman. Westport, Conn.: Greenwood Press: 47–60.

Tuzin, Donald. 1976. *The Ilahita Arapesh: Dimensions of Unity.* Berkeley: University of California Press.

———. 1980. *The Voice of the Tambarian: Truth and Illusion in Ilahita Arapesh Religion.* Berkeley: University of California Press.

Ueda, S., and others. 1989. "Nucleotide Sequences of Immunoglobulin-Epsilon Pseudogenes in Man and Apes and Their Phylogenetic Relationships." *Journal of Molecular Biology* 205: 85–90.

Uehara, Shigeo, and others. 1992. "Characteristics of Predation by the Chimpanzees in the Mahale Mountains National Park, Tanzania." In *Topics in Primatology, Vol. 1: Human Origins.* Tokyo: University of Tokyo Press: 143–158.

Vincent, Anne. 1985. "Plant Foods in Savanna Environments: A Preliminary Report of Tubers Eaten by the Hadza of Northern Tanzania." *World Archaeology* 17: 131–148.

Vrba, Elisabeth S. 1988. "Late Pliocene Climatic Events and Hominid Evolution." In *Evolutionary History of the Robust Australopithecines.* Edited by Frederick E. Grine. New York: Aldine de Gruyter: 405–426.

Walker, Alan. 1981. "Diet and Teeth: Dietary Hypotheses and Human Evolution." *Philosophical Transactions of the Royal Society* 292: 57–64.

Ward, Ingebord L. 1978. "Sexual Behavioral Differentiation: Prenatal Hormonal and Environmental Control." In *Sex Differences in Behavior.* Edited by Richard C. Friedman, Ralph M. Richart, and Raymond L. Vande Wiele. Huntington, N.Y.: Robert E. Krieger: 3–17.

Watson, Catherine. 1994a. "Cry Havoc." *The Independent Magazine* (January 15): 16–20.

———. 1994b. "The Death of Democracy." *Africa Report* (January/February): 26–31.

Watts, David P. 1989. "Infanticide in Mountain Gorillas: New Cases and a Reconsideration of the Evidence." *Ethology* 81: 1–18.

———. 1994. "The Influence of Male Mating Tactics on Habitat Use in Mountain Gorillas (*Gorilla gorilla beringei*)." *Primates* 35:35–48.

———. 1996. "Comparative Socioecology of Gorillas." In *Great Ape Societies*. Edited by William C. McGrew, Toshisada Nishida, and Linda Marchant. Cambridge: Cambridge University Press.

Weil, Liz. 1994. "Revenge of the Girl Next Door." *Boston Magazine* (November): 59ff.

Weiss, Philip. 1995. "Outcasts Digging in for the Apocalypse." *Time* (May 1): 48, 49.

Weller, Sheila. 1994. "Girls in the Gang: A Nineties Nightmare." *Cosmopolitan* (August): 166 ff.

Wendorf, Fred. 1968. "Site 117: A Nubian Paleolithic Graveyard near Jebel Sahara, Sudan." In *The Prehistory of Nubia*, vol. 2. Edited by Fred Wendorf. Dallas: Southern Methodist University Press: 954–995.

Wendorf, Fred, and Romuald Schild. 1986. *The Wadi Kubbaniya Skeleton: A Late Paleolithic Burial from Southern Egypt. The Prehistory of Wadi Kubbaniya*, vol. 1. Edited by Angela E. Close. Dallas: Southern Methodist University Press.

White, Frances J. 1992. "Pygmy Chimpanzee Social Organization: Variation with Party Size and Between Study Sites." *American Journal of Primatology* 26: 203–214.

White, Tim D., Gen Suwa, and Berhane Asfaw. 1994. "*Australopithecus ramidus*, a New Species of Early Hominid from Aramis, Ethiopia." *Nature* 371: 306–312.

———. 1995. "Corrigendum to *Australopithecus ramidus*, a New Species of Early Hominid from Aramis, Ethiopia." *Nature* 375: 88.

Wickler, Wolfgang. 1967. *The Sexual Code: The Social Behavior of Animals and Men.* London: Weidenfeld and Nicolson.

Williams, S. A., and Morris Goodman. 1989. "A Statistical Test that Supports a Human/Chimpanzee Clade Based on Noncoding DNA Sequence Data." *Molecular Biology and Evolution* 6: 325–330.

Wilson, Nanci Koser. 1980. "Styles of Doing Time in a Coed Prison: Masculine and Feminine Alternatives." In *Coed Prison*. Edited by John O. Smykla. New York: Human Sciences: 150–171.

Wilson, T. W. 1986. "Gender Differences in the Inmate Code." *Canadian Journal of Criminology* 28: 297–405.

Wingfield, John C., Carol S. Whaling, and Peter Marler. 1994. "Communication in Vertebrate Aggression and Reproduction: The Role of

Hormones." In *The Physiology of Reproduction*, 2nd ed. Edited by E. Knobil and J. D. Neill. New York: Raven: 303–342.

WoldeGabriel, Giday, and others. 1994. "Ecological and Temporal Placement of Early Pliocene Hominids at Aramis, Ethiopia." *Nature* 371: 330–333.

Wolf, Katherine, and Steve R. Schulman. 1984. "Male Response to 'Stranger' Females as a Function of Female Reproductive Value Among Chimpanzees." *American Naturalist* 123: 163–174.

Wolf, Naomi. 1991. *The Beauty Myth: How Images of Beauty Are Used Against Women.* New York: Doubleday.

———. 1993. *Fire with Fire: The New Female Power and How It Will Change the 21st Century.* New York: Random House.

Wolff, A. 1995. "An Unrivaled Rivalry." *Sports Illustrated* (March 6): 74–84.

Wolfheim, Jaclyn H. 1983. *Primates of the World: Distribution, Abundance, and Conservation.* Seattle: University of Washington Press.

Wolpoff, Milford H. 1980. *Paleoanthropology.* New York: Alfred A. Knopf.

Wood, Bernard. 1994. "The Oldest Hominid Yet." *Nature* 371: 280–281.

Wrangham, Richard W. 1975. *Behavioral Ecology of Chimpanzees in Gombe National Park, Tanzania.* Cambridge: Cambridge University Ph.D. thesis.

———. 1979. "On the Evolution of Ape Social Systems." *Social Science Information* 18: 335–368.

———. 1981. "Drinking Competition in Vervet Monkeys." *Animal Behaviour* 29: 904–910.

———. 1982. "Ecology and Social Relationships in Two Species of Chimpanzee." In *Ecological Aspects of Social Evolution: Birds and Mammals.* Edited by D. I. Rubenstein and Richard W. Wrangham. Princeton: Princeton University Press: 352–378.

———. 1993. "The Evolution of Sexuality in Chimpanzees and Bonobos." *Human Nature* 4: 47–79.

Wrangham, Richard W., Adam P. Clark, and Gilbert Isabirye-Basuta. 1992. "Female Social Relationships and Social Organization of the Kibale Forest Chimpanzees." In *Topics in Primatology, Vol. 1: Human Origins.* Edited by Toshisada Nishida and others. Tokyo: University of Tokyo Press: 81–98.

Wrangham, Richard W., John L. Gittleman, and Colin A. Chapman. 1993. "Constraints on Group Size in Primates and Carnivores:

Population Density and Day-range as Assays of Exploitation Competition." *Behavioral Ecology and Sociobiology* 32: 199–209.

Wrangham, Richard W., William C. McGrew, and Frans B. M. de Waal. 1994. "The Challenge of Behavioral Diversity." In *Chimpanzee Cultures*. Edited by Richard W. Wrangham and others. Cambridge, Mass.: Harvard University Press: 1–18.

Wrangham, Richard W., and others. 1996. "Social Ecology of Kanyawara Chimpanzees: Implications for the THV Hypothesis." In *Great Ape Societies*. Edited by William C. McGrew, Linda F. Marchant, and Toshisada Nishida. Cambridge: Cambridge University Press.

Wrangham, Richard W., and Emily van Zinnicq Bergmann Riss. 1990. "Rates of Predation on Mammals by Gombe Chimpanzees, 1972–1975." *Primates* 31: 157–170.

Wrangham, Richard W., and Barbara B. Smuts. 1980. "Sex Differences in the Behavioural Ecology of Chimpanzees in the Gombe National Park, Tanzania." *Journal of Reproduction and Fertility*, Supplement 28: 13–31.

Wright, Robert. 1994. "Feminists, Meet Mr. Darwin." *The New Republic* (November 28): 34–46.

Yerkes, Robert M. 1925. *Almost Human*. London: Jonathan Cape.

Zihlman, Adriene L., and others. 1978. "Pygmy Chimpanzees as a Possible Prototype for the Common Ancestor of Humans, Chimpanzees and Gorillas." *Nature* (London) 275: 744–746.

Zuckerman, Solly. 1933. *Functional Affinities of Man, Monkeys, and Apes: A Study of the Bearings of Physiology and Behaviour on the Taxonomy and Phylogeny of Lemurs, Monkeys, Apes, and Man*. London: Kegan Paul.

ACKNOWLEDGMENTS

Perhaps the most exciting part of our research was our trip to Zaïre to see bonobos. For help on that journey, we thank several people, including particularly Takayoshi Kano, who provided full and serene hospitality. Dr. Kano allowed us to observe his research and openly shared his observations and conclusions. He and Mrs. Kano shared several pleasant evenings with us at Wamba, while Chie Hashimoto and Evelyn Ono Vineberg took time off from their own pressing research to guide us into the forest. Norbert Likombe Batwafe and Ikenge Justin Lokati also, sensitively and professionally, contributed their time and energy on our behalf; and Karl Ammann was generous and helpful in any number of ways.

In Uganda we are particularly indebted to Linda and Oskar Rothen, Peter Howard, John Kasenene, Lysa Leland, Tom Struhsaker; in the forest: Joseph Basigara, Bart Beerlage, Anja Berle, Joseph Byaruhanga, Lauren Chapman, Nancy Lou Conklin, Kiiza Clement, the late Godfrey Etot, Barbara Gault, Jennifer Gradowski, the late George Kagaba, Christopher Katongole, Elisha Karwani, Samuel Mugume, Francis Mugurusi, Christopher Muruuli, the late Joseph Obua, and Peter Tuhairwe.

Back home, the members of the "Demonic Males" seminar at Harvard in the fall of 1993 helped us examine freshly our thinking on the subject; Nancy Fresco gave a new perspective on girl gangs; while Leah Gardner and Lynnette Simon provided invaluable research and technical assistance.

For inspiration on the intellectual journey, RWW thanks Adam Clark Arcadi, Gilbert Isabirye-Basuta, Rick Bribiescas, Colin Chapman, Richard Connor, Terry Deacon, Peter Ellison, Jane Goodall, David Hamburg, Robert Hinde, Sarah Blaffer Hrdy, Kevin Hunt, Doug Jones, Annette Lanjouw, Mark Leighton, Joseph Manson, Joe Marcus, Jessica Mikszewski, John Mitani, Peggy Novelli, Dan Rubenstein, Maryellen Ruvolo, Barbara Smuts, Shana Swiss, and David Watts. In addition, for comments on the work in progress, we thank Irven DeVore, John Dickson, Tony Goldberg, David Gossman, Nancy Thompson Handler, Marc Hauser, Alison Jolly, Wyn Kelley, Cheryl Knott, Greg Laden, Anne McGuire, David Pilbeam, Lars Rodseth, Elizabeth Ross, Meredith Small, and Rachel Smolker.

For guidance in our expedition into the world of publishing, we are especially grateful to Harry Foster, Lisa Sacks, and Peter Matson. And for making the Kibale chimpanzee study possible, RWW expresses, as always, the strongest appreciation to the Getty Foundation, the L. S. B. Leakey Foundation, the MacArthur Foundation, the National Geographic Society, and the National Science Foundation; also the Department of Zoology of Makerere University and the Government of Uganda, and especially the Ugandan National Research Council, Forestry Department, and the National Parks Board, for permission to work in the Kibale National Park.

INDEX

Adler, Freda, 114
Adolescence
  Mead's study of, 97, 100, 103
  and nature vs. nurture,
    274n.15
Aggression. *See* Male violence;
  Raiding; Violence
Ahlquist, Jon, 29, 39–41
Ajachè, Tata, 110
Alexander the Great, 233
Algerian war of independence,
  women in, 112
Amazon basin, Yanomamö
  among, 64–72, 74, 77, 162, 182
Amazons (warrior women), 109
America, as paradise, 83
Anderson, Charles Robert, 93
Animal observation, revolution-
  ary increase in, 156
Ants, honeypot, 162–64
Apes, 34

African, 38. *See also* Bonobos;
  Chimpanzees; Gorillas
  rainforest, 1, 43, 45, 50–51, 52,
    222–23
  woodland, 31–32, 42–43, 48 (*See
    also* Woodland apes)
Apes, great, 1, 28
  and male violence, 126
  mentality of, 255–57
  patterns of violence among, 131–
    32
  and relationship violence, 152
  See also Bonobos; Chimpanzees;
    Gorillas; Orangutans
Apes-humans relation, 28–30
  and blood-chemistry experi-
    ments, 35–37
  and DNA analysis, 29, 38–41
  and fossil record, 30–31
  and Huxley on humans and go-
    rillas, 32–34

Apes-humans relation (*cont.*)
19th-century skepticism on, 32,
41
and protein structure, 37–38
and chimpanzee as common-
ancestor model, 30, 45–47,
49–50
and convergence, 34–35, 36–37
and divergence date, 42–43
traditional vs. current view of,
29–30, 261
and woodland apes, 31–32, 42,
43, 45, 48
See also Chimpanzees-humans
relation
Arapesh, New Guinea, 279n.16,
281n.16
Arcadia, 83, 106–7
Ardipithecus ramidus, 45
Ardrey, Robert, 11
Aristotle, 28
Arnhem Zoo, Netherlands, 127,
162
Assault
among !Kung, 123
men's vs. women's record of, 113
on Samoa, 105
Augustine, Saint, 213
Australian Aborigines
Buckley's accounts of, 71
male dominance among, 121
warfare mortality rates among
(Murngin), 77
Australopithecines, 31, 31n
See also Woodland apes
Australopithecus afarensis, 31
Australopithecus africanus, 31n
Australopithecus ramidus, 45
Aztecs, 117–18, 161

Baboons, 34
chimpanzee-baboon fight, 179–
80
coordinated fighting of, 130
integration into new troop,
294n.19
and matriotism, 232
as root eaters, 56
teeth of, 178
Bachofen, Johann Jakob, 119
Bangladesh, feminist poet under
death sentence in, 119
Bateson, Gregory, 279n.16
Battering, 26
among chimpanzees, 143–46,
152
See also Wife beating
Batwafe, Norbert Likombe, 215
Beagle, HMS, 94
Behavioral ecology, 22
Benedict, Ruth, 280n.16
Biological determinism
Boas's challenge to, 96–97
of Galton, 95–96
See also Human nature; Nature
vs. nurture
Bipedal travel, 50–51, 61, 269n.3
Boas, Franz, 96–97, 99, 102
and Freeman, 102–3
Boesch, Christophe, 20
Boesch, Hedwige, 20
Bonobos, 1, 26–27, 28, 203–8,
220–21
Chim, 202–3, 220
and chimpanzees, 26–27, 41, 42,
202, 204, 205, 208, 210–12, 215–
16, 222, 225, 293n.9, 293n.10
discovery of, 202
evolution of, 226, 228–29

on evolutionary tree, 261
female bonding among, 208–10,
227, 231, 243
friendliness between communities of, 214–16, 227
geographical range of, 262
Kanzi, 257
male relationships among, 210–
12
and Mongandu, 201–2
monkeys not hunted by, 216–19
party size of, 221–24
and power sharing, 205, 207–8,
238
and lack of rape, 142
sex among, 209–10, 212–14,
215
threats to, 292n.2
See also Females, bonobo
Booee (chimpanzee), 286n.35
Bossou chimpanzee site, 226
Brain, Bob (Charles K.), 56–57
Brain size, 61, 300n.3
evolution of, 256–57
and food, 270n.19
Brewer, Stella, 20
Bridges, Stephen, 248
Brown, Larry, 194
Buckley, William, 71
Buford, Bill, 197–98
Burton, Richard (explorer), 109
Burundi, 2–3

Camp for boys, group-hostility
experiment at, 194–95
Cannibalism
and aggression (Dart), 219
by hyenas, 287n.3
by Marquesans, 88, 90, 92, 93

Chagnon, Napoleon, 64–65, 68,
70
Change in culture. See Cultural
choice
Charismatic megavertebrates,
charting of, 156
Chim (bonobo), 202–3, 220
Chimpanzee Politics (de Waal),
127
Chimpanzees, 1, 28
baboon-chimpanzee fight, 179–
80
and bonobos, 26–27, 41, 42, 202,
204, 205, 208, 210–12, 215–16,
222, 225, 293nn.9,10
conservative evolution of, 46–47
consortships among, 144–45
as demonic male species, 151
on evolutionary tree, 261
females of, 25, 226–27, 242 (See
also Females, chimpanzee)
fighting physiques of, 175
fists as weapons, 180
and Goodall, 7–8, 12, 18, 37,
138, 144, 187
and gorillas, 23, 40, 43–44, 224–
25
and bonobo food supply, 223–24
habitat range of, 262
hand-held weapons of, 180
and infanticide, 151, 159, 166,
226, 289n.22, 297n.17
males' battering of females
among, 143–46, 152
monkeys hunted by, 10–11, 216,
217
mother's support of son among,
206
and origins of patriarchy, 125

Chimpanzees (cont.)
party size among, 165n, 168
as patriarchal, 166
reason in, 187–88
sharing forests with gorillas,
  223–24
social world of, 24
of Tongo rainforest, 57–60
and upright walk, 50–51, 61
violence among
  and differences among species,
    156
  in political struggle, 127–28,
    186–87, 211
  pride as cause of, 190–91
  in raids, 5–7, 12–21, 61–62, 68,
    70–72, 227
  rape, 7, 138, 142, 151
  as xenophobic, 165
Chimpanzees-humans relation,
  23, 25–26, 47–48, 61–62
and battering, 145–46
and bonobos, 26
and common ancestor, 30, 45–
  47, 49–50
and culture, 82
and deliberate killing, 131
in everyday behavior, 8–11, 23–
  24, 29, 49
and evolutionary distance, 29,
  40–42
and female relationships, 25
and male violence, 7, 61, 126
  coalitionary, 24, 233
  human war, 62, 63–64
  raiding, 47, 61–62, 68–72
and other African apes, 23, 38,
  40–41
China
foot binding in, 119

imperial harem of, 233–34
patriarchal empire-building in,
  116–17
Chobe National Park, 160
Choice, cultural. See Cultural
  choice
Clarke, Arthur C., 22
Clausewitz, Karl von, 182
Climatic change
and dispersal of gorillas, 225–26
and evolution of humans, 50, 54,
  60–61, 228–29, 268–69n.1
Clitoridectomy, 119
Coalitionary bonds, 165n
among males, 24–25, 167–68
Coalitionary male violence, 24,
  165n, 231–32, 233, 257
Colobus monkeys, 34
and bonobos, 217–18
chimpanzees' hunting of, 10–11,
  216, 217
Comanche women, and warfare,
  111
Coming of Age in Samoa (Mead),
  100, 106
Control. See Dominance
Convergence, 34–35
and blood-chemistry experiments, 35–37
and molecular level, 36–37
Cost-of-grouping theory, 168–71,
  222
Crime, men-women comparison
  on, 113–15
Crow women, and warfare, 110
Cultural choice, 82
Mead's emphasis on, 100
and remedies for male violence,
  251
and Waorani, 80

Cultural determinism, 106
  and gender differences, 109, 115–16
  and human violence, 176
  and Mead, 106, 279n.16
  See also Nature vs. nurture
Cultural relativism, and Mead, 97
Culture, 82, 128–29
  Boas on, 96–97
  Samoan (Mead's research), 99
  and South Seas paradise, 82, 84
    (*See also* South Pacific)
  and violence, 7, 82
  and war, 69
  See also Nature vs. nurture

Dahomey kingdom female army, 109–10
Damasio, Antonio, 188–90
Dart, Raymond, 31n, 219
Darwin, Charles, 28, 32, 35, 38, 46, 94
  on ingroup-outgroup bias, 196–97
Darwinian selection. *See* Evolution; Natural selection; Sexual selection
Date rape, 137, 141
Dawkins, Richard, 22
Dé (chimpanzee), 12, 16
Decision making, 188–90
Deindividuation, 198
Delaware women, and warfare, 111
Democracy, 245–46
Demonic females, 186n
Demonic males, 151, 167–68
  and intelligence, 258
  as preferred by women, 239–41

unconscious and irrational motivation of, 199
  See also Male violence
Descent of Man and Selection in Relation to Sex, The (Darwin), 32
DeVries, David, 72
de Waal, Frans, 127–28
Djolu village, 4, 200, 201, 263
DNA analysis
  and apes-human relationship, 29, 38–41
  and date of human-chimpanzee divergence, 42
  of gorilla-chimpanzee difference, 44–45
Dolphins, 256–57
Domestic violence. *See* Battering; Wife beating
Dominance
  among chimpanzees, 199, 211
  of males over females, 143–46, 205
  signals of, 211
  and infanticide in gorillas, 150
  of men over women
    and behavior attractive to women, 241
    and gender difference, 277–78n.15
    history of, 251
    and rape, 141–42
    (*See also* Patriarchy)
  rape for purpose of, 141–42
  See also Pride
Dorobo society, 81
du Chaillu, Paul, 33, 43–44, 146–47
Ducks, and rape, 139, 285n.23
Dugum Dani of New Guinea, 77

Ebola virus, 20
Ecological and economic conditions
and cost of grouping, 168–71
and female alliances, 232
and inter-species violence patterns, 70
Elizalde, Manuel, 75–76
Emotional reactions, temperament as, 108n
Emotion-reason relation, 188
Engels, Friedrich, 120, 241
Eskimos, Copper, 81
Eskimos, North Alaskan, 121
Eternal Enemies (Joubert and Joubert), 160
Ethiopia
human ancestor fossils in, 23
as site of woodland ape evolution, 50
Ethnocentrism, 195, 292n.28
Eugenics, 96
European civilization, militaristic aggression of, 116, 117, 118
Evolution
and apes-human relation, 28–30, 34–43, 45–47
traditional vs. current scheme, 29–30, 261
and ape social systems, 232
and benefit vs. cost, 165
of bonobos, 221, 226, 228–29
and comparative anatomy, 34
different speeds of, 46
and error of ancestor resemblance, 46
of human beings, 43, 227–30
and of bonobos, 221, 228–29
continuity in, 61
and "full" humanity, 129, 171

and hominid origins, 268–69n.1
intelligence, 256–58
and status seeking, 191–93
and infanticide, 159–60
and male coalitionary violence, 25–26
of primates, 129–30
from rainforest to woodland, 60–61, 268–69n.1
and bipedal travel, 50–51, 61
and food sources, 51–57
and Tongo chimpanzees, 57–60
and selfish-gene theory, 22–23
sexual selection in, 173–74, 175–76 (*See also* Sexual selection)
violence seen eliminated during, 21
Evolutionary feminism, 124–25, 241–42
Evolutionary viewpoint, unity emphasized in, 250–51

Faben (chimpanzee), 12, 50–51
Females, bonobo, 26, 207–10, 221, 242–43
friendliness initiated by, 215
and male violence, 231–32
and mother-son relationship, 206–7, 221
and parties, 227
Females, chimpanzee, 242
vs. humans, 25
and intercommunity violence, 166–67, 289n.22
and pride, 190
and raiding, 70
relationships among, 226–27
Females, gorilla, 142–43, 242
Females, human
Amazon, 109

vs. chimpanzees, 25
and democracy, 245–47
demonic males preferred by,
239–41
Herland governed by, 236–39
vs. male system, 242–43
mutilation of, 119
and patriarchy, 119, 125
and power seeking, 235
in street gangs, 276–77n.13
Females, hyena, 160, 166, 173–74
dominance battles among, 185
and fighting for food, 155, 170
and infant fratricide, 186
sexual mimicry of, 184–86n
Females, lion, 160
Females, monkey, 232
fighting formation of, 130
Females, orangutan, 133, 242
Females, wolf, 160
Feminism
evolutionary, 124–25, 241–42
and gender differences, 109, 115,
278–79n.15
and patriarchy, 124–25, 241
Figan (chimpanzee), 5, 12
Fisher, Helen, 120
Fists, 180
Food, 51–52
and bonobo temperament, 223–
24
and brain size, 270n.19
gorilla-style, 223, 224
and parties vs. troops, 169–70
and woodland adaptation, 52–
54, 56–61
Football, 248
Football stars, 240–41
Forager societies. See Hunter-
gatherer societies

Forest People (Turnbull), 121–22
Fortune, Reo, 279n.16
Fossey, Dian, 148
Fossils, ape/human, 23, 30–31
Fouts, Roger, 286n.35
Fox women, and warfare, 110–11
Frank, Laurence, 183–84
Frankfurt Zoological Society, 57
Frederick II (Holy Roman Em-
peror), 244n
Freeman, Derek, 102
Friedl, Ernestine, 121
Friedman, Marilyn, 240

Gage, Phineas, 188
Galdikas, Biruté, 136–37
Galton, Francis, 95–96
Galton's Error, 95, 126, 176–77
emotion-reason parallel to, 188
Gauguin, Paul, 11, 84–87, 90, 91,
92, 120
Gender differences
as culturally determined, 109
and patriarchy, 115–16
and domination, 277–78n.15
and traditional feminism, 278–
79n.15
Genetic associations, between
primate species, 37–42
Genetic inheritance, vs. environ-
ment, 95–96. See also Nature
vs. nurture
Gentle Tasaday, The (Nance), 76
Germany under Nazis, and
women in war, 111
Gezo (king of Dahomey), 109
Gibbons, 34
Gigi (chimpanzee), 5, 6
Gilman, Charlotte Perkins, 236–
39

Gingrich, Newt, 106
Godi (chimpanzee), 5–6, 7, 11–12, 16
Goldsmith, Michele, 168–69
Goliath (chimpanzee), 12, 16–17
Gombe National Park, Tanzania, 12
  chimpanzee behavior in, 5, 18–19, 20, 138, 143, 144, 145, 162, 166–67, 180, 216, 217, 226
  on map, 263
  mortality rate from aggression in, 70, 271n.9
  size of chimpanzees from, 203
  and woodlands, 31n
Gone with the Wind (Mitchell), 240
Goodall, Jane, 7–8, 11, 12, 37
  bananas for chimpanzees of, 18
  on battering among chimpanzees, 144
  on rape among chimpanzees, 138
  on reasoned aggression of chimpanzee, 187
  at time of protein experiments, 37
Gorillas, 1, 28, 147–48
  battering absent among, 146
  and chimpanzees, 23, 40, 43–44, 224–25
  and bonobo food supply, 223–24
  conservative evolution of, 46
  as demonic male species, 151
  du Chaillu's experience of, 146–47
  on evolutionary tree, 261
  evolution of habitat of, 224–26
  fighting physiques of, 175
  fists as weapons, 180
  foods of, 223, 224
  habitat range of, 262
  and humans, 29, 32–33, 33–34, 40, 41, 42
  infanticide among, 131, 142, 146–51, 152
  and other African apes, 38
  and rape, 138, 142, 143
  silverback, 142–43, 147
  Gorillas in the Mist (Fossey), 148
Gowaty, Patricia, 124
Great apes. See Apes, great
Greene, Richard Tobias, 89
Group loyalty and hostility, 194–98. See also Intergroup aggression
Gundul (orangutan), 137
Gunung Leuser National Park, Sumatra, 133
Gunung Palung National Park, Borneo, 133

Harems, 233–34
Harmless People, The (Thomas), 76
Hayes, Cathy, 256
Headhunting, modern form of, 271n.1
Henslow, John Stevens, 94
Heredity. See Biological determinism; Human nature; Nature vs. nurture
Herland (Gilman), 236–39
Heyman, Art, 194
Hockey, 248
Hoka-hoka, 209–10, 294n.20
Hölldobler, Bert, 164
Holt, Edward, 99
Hominids
  bipedal travel as hallmark of, 50

dental features of, 52
origins of, 268–69n.1
in traditional view, 29
Howard, Michael, 182, 188, 192
Howler monkeys, 138
Hrdy, Sarah, 124
Hugo (chimpanzee), 6, 12, 179–80
Huli of New Guinea, 77
Human nature
Boas on, 102
Galton on, 95–96
and optimistic images of paradise, 84
and Melville, 90, 93
See also Biological determinism; Nature vs. nurture
Humans
as apes, 233
bodies of, 178, 181
and culture, 82, 128–29 (See also Cultural determinism; Culture)
evolution of, 43, 61, 129, 171, 227–30, 268–69n.1 (See also Evolution)
on evolutionary tree, 261
and gorillas, 29, 32–33, 33–34, 40, 41, 42
mental characteristics of, 182–83
decision making, 188–90
group loyalty and hostility, 194–98
pride, 191–93
(See also Intelligence)
moral aspects of, 229–30
and sex, 212–13
Humans and chimpanzees. See Chimpanzees-humans relation
Humans and great apes. See Apes-humans relation

Human violence
and differences among species, 156
from male-bonded groups, 24–26
questions on, 23, 25–26
and selfish-gene theory, 22–23
See also Intergroup aggression; Male violence; Warfare, human
Humphrey (chimpanzee), 5, 12, 19, 180
Hunter-gatherer (foraging) societies, 74–77
egalitarian ethic of, 74, 109, 120
!Kung San, 76–77, 122–24
Hunting
of monkeys by chimpanzees, 10–11, 216, 217
and murder, 219
Huxley, Thomas, 32–33, 36
Hyenas, spotted, 155–56, 160
in conquered land, 167
dominance battles among, 185
as female-led, 166 (See also Females, hyena)
fratricidal newborns of, 183–84, 186n
hunting party of, 153–55
and male vs. female aggression, 167
and patriotism vs. matriotism, 232
sexual mimicry by females among, 184–86n
and sexual selection, 173–74
as xenophobic, 165

Idani, Gen'ichi, 209, 214
Imperialism, 235–36, 251
Incas, 117
Inclusive fitness theory, 22

India
    Chipko environmental movement in, 246
    suttee tradition of, 119
Infanticide, 155–60
    and chimpanzees, 151, 159, 166, 226, 289n.22, 297n.17
    evolutionary explanation of, 159–60
    among gorillas, 131, 142, 146–51, 152
    among lions, 157–58
Ingroup-outgroup bias, 195–98
Intelligence
    and behavior, 177
    demonic males with, 257–58
    evolution of, 256–58
    and male coalitionary violence, 25
    and relationship violence, 152, 257
    vs. wisdom, 258
Intergroup aggression, 247–51
    as chimpanzee-human similarity, 26, 63
    and party size, 227
    See also Group loyalty and hostility; Raiding; Warfare, human
International relations, and intergroup aggression, 249–51
Irwin, James, 197
Isabirye-Basuta, Gilbert, 20
Islam
    and treatment of women, 119
    and warfare, 118
Israel, women of in military, 112
Italy, 12th-century political systems in, 244, 244–45n
Iten, Oswald, 76
Ituri Forest, kakbas in, 54–56

Janson, Charles, 168–69
Japan, patriarchal and militaristic society in, 117
Jealousy, sexual, 174
Jericho, 171–72
Johanna (ape), 44
Johnson, Samuel, 191
Jomeo (chimpanzee), 12, 291n.19
Joubert, Beverly, 160
Joubert, Dereck, 160

Kabarole, 252–54
Kahama chimpanzee community, 5, 13–18, 19, 21–22, 66, 162
Kakama (chimpanzee), 252, 253
    doll of, 254–55
Kangaroos, 180
Kano, Takayoshi, 204, 206, 208, 215, 221, 223
Kanzi (bonobo), 257
Karagwe, king of, 163
Karwani, Elisha, 255
Kasekela chimpanzee community, 12, 13–18, 19, 66, 162, 166–67, 187
Keith, Sir Arthur, 44
Kibale Forest, Uganda, 20–21, 168, 216, 217, 226, 252, 263
Killer apes, 21, 47
Killer males, females' need for protection from, 159. See also Demonic males; Male violence
Killing of infants. See Infanticide
Killing of same-species adults, 6–7, 63, 155, 160–61, 287n.3
    vs. defeating of rival, 6, 130–31, 155, 161–62
    See also Male violence; Raiding; Violence; Warfare, human

King Kong, 21, 147, 148
Koko (gorilla), 255–56
Kooloo-kamba, 43–44
!Kung San, 76–77, 122–24
Kuroda, Suehisa, 204, 221
Kutai Game Reserve, Borneo, 133, 136

Language, 61, 71–72, 256–57
Langurs, 34
Lanjouw, Annette, 58, 59
Las Casas, Bartolomé de, 117
Lee, Richard, 122
Leni Lanape, 72
Lerner, Gerda, 120, 124
Lilungu site, 216, 218
Linnaeus, Carolus, 266n.3
Lions, 156, 160
  infanticide among, 157–58
  killing of rival among, 160–61
Lokati, Ikenge Justin, 215
Lomako site, 218, 295–96n.38
Lorenz, Konrad, 22, 156, 219
"Lucy" (fossil), 31, 45
Luit (chimpanzee), 127–28, 162
Lynchings, 197

Macaques, 34
  Barbary, 28
  rhesus, 130, 232
MacKinnon, John, 135
McVeigh, Timothy, 247
Madam Bee (chimpanzee), 289n.22
Mae Enga of New Guinea, 73–74, 77
Mafuca (ape), 44
Mahale Mountains National Park, 19, 20, 166, 216, 217, 263
Majuro women, and warfare, 110

Malaysia, Semai Senoi of, 81
Male-bonded patrilineal kin groups, 24
Males, and visions of paradise, 84, 87, 91, 92–93
  and Mead's research, 99, 103
Male violence (aggression)
  archaeological record on, 172, 290n.33
  coalitionary, 24, 165n, 231–32, 233, 257
  and differences among species, 126
  vs. examples of female-led societies, 167
  and female preferences, 239–41
  and Herland, 238
  and human body, 172, 178–79, 181–82
  as human universal, 126
  long history of, 249, 251
  of Melville's Typees, 93
  from party-gang traits, 167
  and sexual selection, 173–74
  and team sports, 248
  and temperament, 108–9
  and crimes, 113–15
  dominance as goal of, 199
  and examples of female warriors, 109–13
  and patriarchy, 114, 115–16, 125 (See also Patriarchy)
  and reproductive interests, 235
  unconscious and irrational basis of, 199, 249
  See also Relationship violence; Violence; Warfare, human
Maori women, and warfare, 110
Mao Tse-tung, 243

Marcos, Ferdinand, and Tasaday, 76

Margaret Mead and Samoa (Freeman), 102

Marital rape, 141

Marquesas Islands, 87–88
and Darwin, 94
and Gauguin, 86–87
and Melville, 87, 89

Marriage, as problem for women, 243

Masai Mara Game Reserve, Kenya, 184

Masculinizing
behavioral effects of, 185
of female hyenas, 184n, 185–86n

Matama, Hillali, 5

Matriarchy, 119–20

Matrilineal/matrilocal communities, aggressive coalitions in, 24–25

Matriotism, 232
in Herland, 238–39

Mayas, 117

Mbuti of Central Africa, 121–22

Mead, Margaret, 84, 94, 97–106, 115, 120, 279n.16
in New Guinea, 279–81n.16

Meat-eating
and evolutionary ancestors, 61
and rainforest-woodland evolution, 53–54
See also Hunting

Melville, Herman, 84, 87–94, 98, 120

Melville, Thomas, 89

Melville in the South Seas (Anderson), 93

Mendel, Gregor, 96

Michigan Militia, 247

Mike (chimpanzee), 12, 187

Mitani, John, 136

Mitchell, Margaret, 240

Mob behavior, 197–98

Moe, Doug, 194

Mole rats, 57

Mongandu, 201, 215

Monkeys, 34
Barbary macaques, 28
colobus, 10–11, 34, 216, 217–18
howler, 138
infanticide among, 156–57
as matriotic, 232
muriqui, 174–76, 290n.2, 295n.35
rhesus macaques, 130, 232
vervet, 59, 130

Mother-son relationship, among bonobos, 206–7, 221

Mt. Assirik, 31n

Mundugumor, New Guinea, 279n.16

Murder
and hunting, 219
on Samoa, 105

Muriqui, 174–76, 290n.2, 295n.35

Murngin Aborigines of Australia, 77

Murumba, Ngoga, 4

Nance, John, 76

Nasrin, Taslima, 119

Native Americans
and honeypot ants, 163
and surprise in warfare, 72–73
women in wars of, 110–11

Naturalistic fallacy, 284n.6

Natural selection
selfish-gene theory of, 22–23
violence seen eliminated by, 21

See also Evolution; Sexual selection
Nature vs. nurture, 95
  and Boas, 97
  Galton's Error on, 95, 126, 177
  and male violence, 126
  and Mead, 106
  See also Biological determinism; Cultural determinism; Human nature
Navaho women, and warfare, 110
Ndadaye, Melchior, 2
New Guinea
  battered wives in, 119
  Dugum Dani warfare mortality rates in, 77
  Huli violent mortality rate in, 77
  Mae Enga warfare in, 73–74, 77
  Mead's research in, 279–81n.16
Nikkie (chimpanzee), 128, 162
1984 (Orwell), 249–50
Niokola-Koba National Park, Senegal, 19–20, 263
Nisa: The Life and Words of a !Kung Woman (Shostak), 123–24
Nishida, Toshisada, 19
Nunki (gorilla), 150
Nuttall, George, 35, 37

Observation of animals, revolutionary increase in, 156
Oklahoma City bombing, 247
Olson, Norman, 247
On Aggression (Lorenz), 22
Orangutans, 1, 28, 132–33
  big vs. small males, 134–35
  conservative evolution of, 46
  and cost-of-grouping theory, 171

as demonic male species, 151
divergence of from African apes, 42
on evolutionary tree, 261
female, 133, 242
fighting physiques of, 175
and humans, 38
rape among, 131, 132–38, 139–43, 152
as solitary and aggressive, 193
Orgasm, in female primates, 294n.20
Original sin, 22
Origin of Species (Darwin), 94
Orokaiva women, and warfare, 110
Orwell, George, 249–50

Pacifist species. *See* Bonobos; Muriqui
Pago Pago, Mead at, 97, 98. *See also* Samoa
Pakistan, mortality rate for girls in, 119
Palmer, Craig, 138
Paradise, 82
  Arcadia as, 83, 106–7
  chimpanzee life seen as, 11
  as cultural theme, 83
  South Pacific as, 83–94, 97–106
  unreality of, 107
Party(ies), 13n, 165n
  of bonobos, 221–24, 227
Party-gang species, 165–68
  and cost-of-grouping theory, 168, 169–71
  and foreign policy, 235
  and group hostility, 198
Passion (chimpanzee), 226, 297n.17

Patriarchy
  in chimpanzee society, 166
  and contemporary preagricul-
    tural societies, 121–24
  and cultural theories of gender,
    115–16
  as flexible strategy, 125
  and search for matriarchy, 119–
    20
  and universality of patriarchy,
    125
  Engels on, 120, 241
  and evolutionary feminists, 124–
    25, 241–42
  Herland as freedom from, 237
  institutionalization of (Lerner),
    120
  and male violent crime, 114
  and militaristic dominance, 116–
    18
  and non-Western societies, 118–
    19
  origins of, 125, 232–33
Patrilineal kin groups, male-
  bonded, 24
Patriotism, 231
  and aggression, 233
Patterson, Francine, 255
Peloponnesian War, 192–93
Political power, 243
  personalized and institutional-
    ized, 244–47
  and single-sex institutions,
    299n.7
Pom (chimpanzee), 226, 297n.17
Porter, David, 88, 89
Porter, Roy, 116
Poussin, Nicolas, 107
Pride, 190
  and human competition, 191–93

  and human warfare, 192
  and male chimpanzees, 190–91
  See also Dominance
Primates, 32, 129
  aggressive behavior of, 130–31
  early evolution of, 129–30
  humans as, 32–34
Primitive War (Turney-High), 72
Private property, and subordina-
  tion of women (Engels), 120
Putnam, Robert D., 244
Pygmies, in Ituri Forest, 55–56
Pygmy chimpanzee, 26, 202, 292–
  93n.5. See also Bonobos

Queen of Beasts (film), 158
Quichua, and Waorani, 78–79

Raiding
  and bonobos, 221, 227
  among chimpanzees, 5–7, 12–
    20, 47, 61–62, 68, 227
  and hyena attacks, 160
  as imperialist, 235
  by !Kung San, 76
  among party-gang species, 165
  vs. territorial fights, 130–31
  among Waorani, 78
  among Yanomamö, 67–68, 71–72
  See also Warfare, human
Rainforest, 46
  fruit for primates in, 46
  root starches meager in, 55
Rainforest apes, 1, 43, 45
  foods needed by, 52
  forests as food reservoirs for,
    222–23
  woodland adaptation of, 50–51
Range, 14n
Rape, 26

among chimpanzees, 7, 138, 142, 151
date rape, 137, 141
definition and frequency of, 282–83n.42
dominance as purpose of, 141–42
among Eskimos, 121
and evolutionary feminists, 125
evolutionary role of, 132, 138–39, 140, 285n.23
and Gingrich's optimism, 106
and Judeo-Christian patriarchy, 116
and !Kung San, 122
marital, 141
and Mbuti, 122
and Mead on Samoa, 101, 104
men's vs. women's record of, 113
by militarist patriarchies, 117
among orangutans, 131, 132–38, 139–43, 152
and primates, 138
and social system, 142
and Tahitians, 86
of woman by orangutan, 137–38
Reason-emotion relation, 188
Reflections of Eden (Galdikas), 136
Relationship violence
chimpanzee battering as, 146
and intelligence, 152, 257
women's role in, 243
See also Battering; Infanticide; Rape; Sexual violence
Reproduction as evolutionary factor, 165
and infanticide, 159–60
and patriarchy, 125
and power striving, 233–35

and preference for demonic males, 239
and rape, 138–41
and status seeking, 191
See also Sexual selection
Rijksen, Herman, 136, 140
Robarchek, Carole, 79, 80
Robarchek, Clayton, 79, 80
Robber's Cave experiment on group hostility, 194–95
Roman Empire, slave children in, 234
Roots
and Tongo chimpanzees, 59
as woodland food source, 54–57, 59–60
Rousseau, Jean Jacques, 11, 84
Ruwenzori (chimpanzee), 20
Rwanda, 3–4, 58

Samoa, 97–98
Freeman on, 103
Mead's study of, 97–102, 103–6
Sanday, Peggy Reeves, 121–22
Schwarz, Ernst, 202
Schwarzenegger, Arnold, 240
Scorpion flies, 138–39
Selfish Gene, The (Dawkins), 22
Selfish-gene theory of natural selection, 22–23
Semai Senoi, 81
Serengeti National Park, Tanzania, 157–58
Seville Statement on Violence, 176
Sex
among bonobos, 213–14, 221, 295n.31
between females, 209–10, 215
among chimpanzees, 11

Sex (cont.)
  for Gauguin, 86, 87
  and humans, 212–13
  and Mead on Samoa, 100–101,
    103
  among muriquis, 174
  among orangutans, 133
Sex and Temperament (Mead),
  280n.16
Sexual coercion hypothesis, 141–
  42
Sexual mimicry, among hyenas,
  184–86n
Sexual selection, 173–74, 175–
  76
  and human aggression, 183
  and human rulers, 234
  and male temperament, 235
  and male upper-body strength,
    181–82
  and patriotism, 233
  and pride, 192
Sexual violence
  among chimpanzees, 7
  and patriarchy (Porter), 116
  See also Rape; Relationship vio-
    lence
Shakur, Sanyika, 193
Shostak, Marjorie, 122, 123
Sibley, Charles, 29, 39–41
Simmel, Georg, 198
Simon, Rita, 114
Single-sex institutions, 299n.7
Slim (chimpanzee), 291n.19
Small, Meredith, 124
Smuts, Barbara, 124, 141, 146
Smuts, Robert, 141, 146
Sniff (chimpanzee), 12, 17
Social system(s), 152
  of bonobos, 205

evolutionary logic of, 232
and rape, 142
as variable (human), 82
See also Culture; Patriarchy; Po-
  litical power
Sociobiology, 22
South Pacific
  Darwin on, 94
  paradise envisioned in, 83–84
    by Gauguin, 84–87
    by Mead, 97–106
    by Melville, 87–94
Soviet Union, women of in war,
  111–12
Speke, John Hanning, 163
Sperm competition, 175, 176n
Spock, Benjamin, 105
Spotted hyenas. See Hyenas,
  spotted
Stewart, Charles, 89
Street gangs, 193, 276–77n.13
Switzerland, 80–81

Tahiti
  Darwin on, 94
  and Gauguin, 84–86
  and Melville, 87, 89–90
Taï Forest of Ivory Coast, 20, 21,
  216, 217, 218, 263
Tanjung Puting National Park,
  Borneo, 132–33, 136
Tarzan, 21
Tasaday, 75–76
Taung baby, 31n
Taylor, Helen, 240
Tchambuli culture, 279n.16
Technology, and intergroup ag-
  gression, 249
Teeth
  bonobo, 224

and human fighting ability, 178–79
and woodland adaptation, 52–53, 57
Temperament, 108n
and behavior, 177
and domination, 199
gender differences in, 115–16
and intelligence, 258
male, 108–16, 125, 199, 235
and violence, 108–9
Territorial fights, 286n.2
by honeypot ants, 163
among primates, 130–31
chimpanzees, 20
Territory, 14n
Testosterone, 181, 239
Thomas, Elizabeth Marshall, 76
Thucydides, 192
Tiger (gorilla), 150
Tikopia society, 81
Toda society, 81
Tongo rainforest, 57–60, 61
Tonkinson, Robert, 121
Tools
of chimpanzees, 180
and evolutionary ancestors, 61
for root digging, 56–57
Trochmann, David, 248
Troop, 13n
and cost-of-grouping theory, 168–69
Tuhairwe, Peter, 255
Turnbull, Colin, 121–22
Turney-High, Harry, 72
Twin studies, and Galton on nature vs. nurture, 96
Typee (Melville), 87, 90–93
Typees, 88–90, 91

Ugalla, Tanzania, 31n
United Nations, 250
Michigan Militia view of, 247
Upper-body strength, 181
Uwilingiyimana, Agathe, 3

Vervet monkeys, 59, 130
Viki (chimpanzee), 256
Violence (aggression)
and Amazons, 109
and bonobos, 220–21
among chimpanzees, 5–7, 12–20, 127–28, 156, 227 (See also under Chimpanzees)
and culture, 7, 82
and evolutionary feminists, 125
human, 22–26, 156
among party-gang species, 165–68
and unbalanced power, 162
among female hyenas, 185
among honeypot ants, 164
as uniquely human, 6–7
See also Infanticide; Killing of same-species adults; Male violence; Relationship violence; Warfare, human
Virgil
on Arcadia, 83
and search for primitive, 120
Virunga Volcanoes, 149
Visit to the South Seas, A (Stewart), 89

Wamba village and bonobo site, 201, 205, 214, 218
Waorani, 78–80
Warfare, human, 22, 63
and anthropological/literary record

Warfare, human (cont.)
   hunter-gatherer peoples, 71, 74–77
   Mae Enga of New Guinea, 73–74
   in Marquesas Islands, 88
   and Mead on Samoa, 101, 104–5
   and mortality statistics for independent peoples, 77
   Semai Senoi, 81
   societies without military organization, 81
   Typees, 91
   Waorani, 78–80
   Yanomamö of Amazon, 64–72, 74, 77, 162, 182
   archaeological record of (Jericho), 171–72
   and chimpanzee violence, 62, 63–64
   as common, 81–82
   and cultural determinism, 176
   cultural dimensions of, 69, 271n.1, 276n.4
   differing benefits from, 167
   and honeypot ants, 162–64
   and modern vs. tribal life, 77–78
   origin of, 22
   from party-gang traits, 167
   pride as cause of, 192
   primitive vs. "civilized," 72
   and reasoned decisions, 182–83
   superior force as principle of, 162
   surprise as element of, 72–73
   and Switzerland, 80–81
   women participants in, 109–13
   and world government, 250
   See also Intergroup aggression; Raiding

Wayward Servants (Turnbull), 121–22
Weapons
   of chimpanzees, 180
   of woodland apes, 180, 181–82
Weil, Liz, 114
Western civilization, militaristic aggression of, 116, 117, 118
Wife beating, 119
   among !Kung, 123–24
   among Mbuti, 121
Williams, John, 104–5
Wisdom, 258
Wolf, Naomi, 114, 247
Wolves, 160, 167
Women. See Females, human
Women's movement, and female crime, 114
Woodland apes, 31–32, 42–43, 48
   evolutionary branchings of, 61
   and foraging parties, 229
   human evolution from, 227
   and primate evolution, 129
   size of, 45
   teeth of, 179
   weapons of, 180, 181–82
World government, 250

Yanomamö, warfare among, 64–72, 74, 77, 162, 182
Yerkes, Robert, 202–3, 220
Yeröen (chimpanzee), 128, 162

Zaïre River, bonobos and chimpanzees separated by, 222, 223, 224–25
Zulus, 117